Introduction

Chapter 1: Soil Health and Regeneration

Chapter 2: Water Management and Conservation

Chapter 3: Biodiversity Enhancement

Chapter 4: Carbon Sequestration and Climate Mitigation

Chapter 5: Sustainable Livestock Management

Chapter 6: Agroforestry Systems

Chapter 7: Urban Agriculture and NbS

Chapter 8: Policy and Economic Incentives

Chapter 9: Farmer and Community Engagement

Chapter 10: Future Directions and Innovations

Conclusion

Introduction

Nature-based solutions (NbS) are innovative strategies that harness the power of nature to address societal challenges, such as climate change, food security, water scarcity, and biodiversity loss. These solutions involve the sustainable management and use of natural resources and ecosystems to improve environmental health, enhance resilience, and support human well-being. NbS include a range of practices, from reforestation and wetland restoration to agroecology and sustainable water management.

The importance of NbS lies in their ability to provide multiple benefits simultaneously. Unlike conventional approaches that often focus on a single issue, NbS offer integrated solutions that can address various challenges in a holistic manner. By working with nature, these solutions promote biodiversity, improve ecosystem services, and contribute to climate change mitigation and adaptation. Moreover, NbS are often more cost-effective and sustainable in the long term compared to traditional engineering solutions.

Overview of Sustainable Agriculture

Sustainable agriculture is a farming approach that seeks to produce food, fiber, and other agricultural products in a way that preserves and enhances environmental quality, supports economic viability, and promotes social equity. It involves practices that maintain soil health, conserve water, reduce chemical inputs, and enhance biodiversity. Sustainable agriculture aims to meet the needs of the present without compromising the ability of future generations to meet their own needs.

Key principles of sustainable agriculture include:

1. Ecological Balance: Maintaining and enhancing the natural processes and cycles in agricultural systems, such as nutrient cycling, soil formation, and water regulation.

2. Resource Efficiency: Using resources such as water, energy, and nutrients more efficiently and reducing waste and emissions.

3. Diversity: Promoting genetic, species, and landscape diversity to enhance resilience and reduce vulnerability to pests, diseases, and climate change.

4. Resilience: Building the capacity of agricultural systems to withstand and recover from adverse events, such as droughts, floods, and pest outbreaks.

5. Social Equity: Ensuring fair access to resources, opportunities, and benefits for all stakeholders, including smallholder farmers, women, and marginalized communities.

Sustainable agriculture is essential for ensuring food security, reducing environmental impacts, and supporting rural livelihoods. It aligns with the broader goals of sustainable development and contributes to the achievement of the United Nations Sustainable Development Goals (SDGs).

The Role of NbS in Addressing Environmental Challenges

Environmental challenges such as climate change, land degradation, water scarcity, and biodiversity loss are increasingly threatening the sustainability of agricultural systems and the well-being of communities worldwide. NbS offer promising solutions to these challenges by leveraging natural processes and ecosystems to enhance resilience and sustainability.

1. Climate Change Mitigation and Adaptation:

- Mitigation: NbS can sequester carbon in soils and vegetation, reducing greenhouse gas (GHG) emissions and mitigating climate change. Practices such as reforestation, agroforestry, and cover cropping increase carbon storage and improve soil health.

- Adaptation: NbS enhance the adaptive capacity of agricultural systems to cope with climate variability and extremes. For example, agroforestry systems provide shade, windbreaks, and improved microclimates, helping crops and livestock withstand heatwaves and droughts.

2. Soil Health and Erosion Control:

- NbS improve soil structure, fertility, and water retention, reducing erosion and degradation. Techniques such as no-till farming, cover cropping, and organic amendments enhance soil organic matter and microbial activity, promoting healthy soils that are more resilient to erosion and nutrient loss.

3. Water Management and Conservation:

- NbS enhance water availability and quality by improving watershed health and reducing runoff and pollution. Practices such as wetland restoration, riparian buffers, and rainwater harvesting help capture, store, and filter water, ensuring a reliable supply for agricultural and domestic use.

4. Biodiversity Conservation:

- NbS promote biodiversity by creating habitats and supporting diverse species. Agroecological practices, such as polycultures, hedgerows, and integrated pest management (IPM), enhance on-farm biodiversity and ecological interactions, reducing the need for chemical inputs and improving ecosystem services.

5. Resilience to Environmental Stressors:

- NbS build resilience to environmental stressors by diversifying production systems and enhancing ecological functions. Mixed farming systems, agroforestry, and rotational grazing distribute risks and improve the capacity of agricultural landscapes to recover from disturbances.

Scope and Objectives of the Book

This book explores the potential of nature-based solutions to transform agricultural systems into more sustainable and resilient ones. It provides a comprehensive overview of various NbS practices and their applications in different contexts, highlighting their benefits, challenges, and implementation strategies.

The primary objectives of this book are:

1. To Define and Explain NbS:

- Provide a clear and detailed understanding of what NbS are and how they function in the context of sustainable agriculture.

2. To Explore the Benefits of NbS:

- Highlight the multiple benefits of NbS for environmental health, agricultural productivity, climate resilience, and social well-being.

3. To Present Practical Applications:

- Offer practical guidance on how to implement NbS in various agricultural settings, from smallholder farms to large-scale operations.

4. To Address Challenges and Barriers:

- Discuss the challenges and barriers to the adoption of NbS, including technical, economic, and social factors, and propose solutions to overcome them.

5. To Promote Policy and Economic Support:

- Advocate for supportive policies and economic incentives that encourage the adoption of NbS and sustainable agricultural practices.

6. To Engage Stakeholders:

- Emphasize the importance of engaging farmers, communities, policymakers, and other stakeholders in the planning, implementation, and monitoring of NbS.

7. To Look to the Future:

- Explore emerging trends, innovations, and future directions for NbS in sustainable agriculture, and envision a path forward for scaling up and mainstreaming these solutions.

Through this book, readers will gain a deeper understanding of the vital role that nature-based solutions can play in creating a more sustainable and resilient agricultural future. The book aims to inspire and equip farmers, researchers, policymakers, and practitioners with the knowledge and tools needed to implement NbS effectively and contribute to the global movement towards sustainable development.

Chapter 1: Soil Health and Regeneration

Soil health is the foundation of sustainable agriculture. It plays a critical role in determining the productivity and resilience of agricultural systems. Healthy soil supports plant growth, enhances water retention, improves nutrient cycling, and helps sequester carbon, contributing to climate change mitigation. However, modern agricultural practices have often led to soil degradation, threatening food security and ecosystem health.

This chapter delves into the importance of soil health and the current challenges in soil management. It then explores various nature-based soil regeneration techniques, such as cover cropping, crop rotation, composting, and no-till farming. These practices offer sustainable solutions to restore and maintain soil health, ensuring long-term agricultural productivity and environmental sustainability.

Understanding Soil Health

Soil health is a cornerstone of sustainable agriculture, influencing the productivity, resilience, and environmental impact of farming systems. Healthy soils are essential for growing crops, supporting biodiversity, managing water, and mitigating climate change. This section explores the importance of soil health and the current challenges in soil management, providing a foundation for understanding how nature-based solutions can restore and enhance soil vitality.

Importance of Soil Health

Soil health, often referred to as soil quality, encompasses the physical, chemical, and biological properties that make soil capable of sustaining plant and animal productivity, maintaining environmental quality, and promoting plant and animal health. Healthy soil is rich in organic matter, teeming with microorganisms, and has a balanced structure that supports water infiltration and root growth.

Nutrient Cycling and Plant Growth

Healthy soils are critical for nutrient cycling, which is the process of converting essential elements such as nitrogen, phosphorus, and potassium into forms that plants can absorb. Microorganisms in the soil, including bacteria and fungi, play a vital role in breaking down organic matter and releasing nutrients. This continuous recycling of nutrients is fundamental for plant growth and productivity.

Water Retention and Drainage

Soil health significantly affects water retention and drainage. Soils with good structure and high organic matter content can retain moisture, reducing the need for irrigation and helping plants survive during dry periods. Additionally, well-structured soils allow excess water to drain away, preventing waterlogging and root diseases.

Biodiversity and Ecosystem Services

Healthy soils support a diverse array of organisms, from microorganisms like bacteria and fungi to larger creatures such as earthworms and insects. This biodiversity is crucial for ecosystem services, including decomposition, pest control, and disease suppression. The interactions among soil organisms also contribute to soil structure and nutrient availability.

Carbon Sequestration and Climate Change Mitigation

Soils act as significant carbon sinks, sequestering carbon from the atmosphere and helping to mitigate climate change. Practices that enhance soil health, such as adding organic matter and reducing tillage, increase soil carbon storage. This not only helps in fighting climate change but also improves soil fertility and structure.

Resistance to Erosion and Degradation

Healthy soils are more resistant to erosion and degradation. Good soil structure, high organic matter content, and robust plant cover protect soil from being washed or blown away by wind and water. This resistance is essential for maintaining long-term agricultural productivity and environmental health.

In summary, the importance of soil health cannot be overstated. It underpins the ability of agricultural systems to produce food, support biodiversity, manage water resources, and mitigate climate change. Understanding and maintaining soil health is therefore fundamental to sustainable agriculture and environmental stewardship.

Current Challenges in Soil Management

Despite its importance, soil health is under threat from various human activities and environmental pressures. Modern agricultural practices, climate change, and land-use changes have led to widespread soil degradation. Addressing these challenges is critical for ensuring the sustainability of agricultural systems and the ecosystems they support.

Soil Erosion

Soil erosion is one of the most significant challenges in soil management. It occurs when soil is removed by wind, water, or human activity faster than it can be naturally replenished. This process depletes the topsoil, which is rich in nutrients and organic matter, reducing soil fertility and productivity. Practices such as deforestation, overgrazing, and improper tillage exacerbate soil erosion.

Loss of Soil Organic Matter

Soil organic matter is vital for soil health, providing nutrients, improving soil structure, and supporting microbial activity. However, conventional farming practices, such as continuous monocropping and excessive use of chemical fertilizers, deplete soil organic matter. This loss reduces soil fertility, water retention, and resilience to erosion and compaction.

Soil Compaction

Soil compaction occurs when soil particles are pressed together, reducing pore space and limiting the movement of air, water, and roots. It is often caused by heavy machinery, overgrazing, and repeated foot traffic. Compacted soils have poor drainage, limited root growth, and reduced microbial activity, leading to decreased crop yields and increased vulnerability to erosion.

Nutrient Depletion and Imbalance

Intensive agriculture can lead to nutrient depletion and imbalance in soils. Continuous cropping without adequate replenishment of nutrients exhausts soil fertility. Over-reliance on synthetic fertilizers can also disrupt nutrient balance, leading to deficiencies or toxicities that harm plant growth and soil health. This imbalance affects crop productivity and soil ecosystem functioning.

Contamination and Pollution

Soils can become contaminated with harmful substances such as heavy metals, pesticides, and industrial pollutants. These contaminants can enter the soil through agricultural runoff, industrial emissions, and improper waste disposal. Soil contamination poses risks to plant and animal health, reduces soil fertility, and can lead to bioaccumulation of toxins in the food chain.

Climate Change Impacts

Climate change exacerbates many soil management challenges. Increased frequency and intensity of extreme weather events, such as droughts and heavy rainfall, accelerate soil erosion and degradation. Changes in temperature and precipitation patterns also affect soil moisture, microbial activity, and organic matter decomposition.

Adapting soil management practices to cope with climate change is essential for maintaining soil health.

Land Use Changes

Conversion of natural landscapes to agricultural or urban areas disrupts soil ecosystems and reduces soil health. Deforestation, wetland drainage, and urban sprawl remove vegetation cover, alter soil structure, and reduce biodiversity. Sustainable land-use planning and conservation practices are needed to protect soil resources and maintain ecosystem services.

In conclusion, soil management faces numerous challenges that threaten soil health and agricultural sustainability. Addressing these challenges requires a shift towards more sustainable practices that enhance soil health, protect against erosion and degradation, and adapt to changing environmental conditions. Understanding the current challenges in soil management is the first step towards implementing effective nature-based solutions that can restore and maintain healthy soils.

Nature-Based Soil Regeneration Techniques

Nature-based soil regeneration techniques are essential for restoring and maintaining healthy soils. These methods leverage natural processes to enhance soil fertility, structure, and biodiversity, ensuring long-term agricultural productivity and environmental sustainability. This section explores four key techniques: cover cropping, crop rotation, composting, and no-till farming.

Cover Cropping

Cover cropping involves growing specific plants, known as cover crops, primarily for the benefit of the soil rather than for crop production. These crops are usually grown during the off-season when main crops are not being cultivated, providing a living cover that protects and improves the soil.

Soil Protection and Erosion Control

Cover crops protect the soil from erosion by wind and water. Their roots bind the soil particles together, reducing the risk of soil being washed or blown away. This is particularly important on sloped land or during periods of heavy rainfall.

Nutrient Enhancement

Cover crops can enhance soil nutrients in several ways. Leguminous cover crops, such as clover and vetch, fix atmospheric nitrogen into the soil, increasing its fertility. Other cover crops, like radishes and mustard, can capture and recycle nutrients from deep soil layers, making them available for subsequent crops.

Organic Matter and Soil Structure

As cover crops grow and decompose, they add organic matter to the soil. This organic matter improves soil structure, increasing its ability to retain water and nutrients. Improved soil structure also enhances root penetration and microbial activity, fostering a healthier soil ecosystem.

Weed Suppression and Pest Control

Dense cover crop stands can suppress weed growth by outcompeting them for light, water, and nutrients. Additionally, some cover crops release natural chemicals that inhibit weed germination. Cover crops can also break pest and disease cycles by disrupting the habitat of specific pests.

Cover cropping is a versatile and effective technique for improving soil health. It provides multiple benefits, including soil protection, nutrient enhancement, and pest control, making it an integral part of sustainable agricultural practices.

Crop Rotation

Crop rotation involves growing different types of crops in a planned sequence on the same field. This practice helps maintain soil fertility, reduce pest and disease pressure, and improve overall soil health.

Soil Fertility and Nutrient Management

Different crops have varying nutrient requirements and contributions. Rotating crops with different nutrient demands prevents the depletion of specific soil nutrients. For example, a legume crop that fixes nitrogen can be followed by a nitrogen-demanding cereal crop, balancing soil nutrient levels.

Pest and Disease Management

Crop rotation disrupts the life cycles of pests and diseases that are specific to certain crops. By changing the crop species, the pests and pathogens lose their food source, reducing their populations. This decreases the need for chemical pesticides and promotes a healthier crop environment.

Improved Soil Structure and Organic Matter

Different crops have different root structures and growth habits, which can benefit soil structure. Deep-rooted crops break up compacted soil layers and improve soil aeration, while shallow-rooted crops protect the soil surface. Rotating crops also adds a variety of organic residues to the soil, enhancing organic matter content.

Weed Suppression

Rotating crops with different growth habits and planting schedules can suppress weed populations. For instance, a densely planted cover crop can outcompete weeds, while a crop with a different growing season can break the weed's life cycle. This reduces the reliance on herbicides and promotes more sustainable weed management.

Crop rotation is a time-tested method for maintaining soil health and productivity. By varying the types of crops grown, farmers can enhance soil fertility, manage pests and diseases, and improve soil structure, leading to more sustainable agricultural systems.

Composting

Composting is the process of decomposing organic matter, such as crop residues, animal manure, and food waste, into a nutrient-rich soil amendment known as compost. This technique recycles organic waste into valuable soil-enhancing material.

Nutrient-Rich Amendment

Compost is rich in essential nutrients, including nitrogen, phosphorus, and potassium, which are vital for plant growth. Adding compost to soil improves its nutrient content, reducing the need for synthetic fertilizers and enhancing soil fertility naturally.

Soil Structure and Moisture Retention

Compost improves soil structure by increasing its organic matter content. This enhances the soil's ability to retain moisture and nutrients, making it more resilient to drought conditions. Improved soil structure also facilitates root growth and microbial activity, promoting a healthy soil ecosystem.

Microbial Activity and Soil Health

Composting introduces beneficial microorganisms to the soil, which play a crucial role in nutrient cycling and organic matter decomposition. These microorganisms enhance soil health by breaking down organic matter, suppressing soil-borne diseases, and improving soil structure.

Waste Reduction and Environmental Benefits

Composting helps reduce the amount of organic waste sent to landfills, reducing greenhouse gas emissions and environmental pollution. It also recycles nutrients within the farm system, promoting a closed-loop approach to nutrient management.

Composting is an effective way to enhance soil health and fertility. By recycling organic waste into valuable soil amendments, composting supports sustainable agriculture and reduces the environmental impact of farming practices.

No-Till Farming

No-till farming, also known as zero tillage, is an agricultural practice where crops are grown without disturbing the soil through traditional tillage methods. This technique maintains soil structure and reduces erosion, enhancing soil health and sustainability.

Soil Structure and Organic Matter

No-till farming preserves soil structure by minimizing soil disturbance. This maintains soil aggregates, which are crucial for water infiltration and root growth. Additionally, organic matter remains on the soil surface, providing a habitat for beneficial microorganisms and protecting the soil from erosion.

Water Conservation and Erosion Control

By leaving crop residues on the soil surface, no-till farming enhances water infiltration and reduces runoff. This conserves soil moisture and prevents erosion, protecting the topsoil and maintaining soil fertility.

Reduced Fuel and Labor Costs

No-till farming reduces the need for mechanical tillage, lowering fuel and labor costs. This makes it an economically viable option for farmers, while also reducing greenhouse gas emissions associated with tillage operations.

Carbon Sequestration and Climate Mitigation

No-till farming increases soil organic carbon levels by minimizing soil disturbance and enhancing organic matter retention. This contributes to carbon sequestration and helps mitigate climate change by storing carbon in the soil.

No-till farming is a sustainable practice that enhances soil health, conserves water, and reduces environmental impacts. By preserving soil structure and organic matter, no-till farming supports resilient agricultural systems and promotes long-term sustainability.

In conclusion, nature-based soil regeneration techniques such as cover cropping, crop rotation, composting, and no-till farming are essential for restoring and maintaining healthy soils. These practices leverage natural processes to enhance soil fertility, structure, and biodiversity, ensuring the sustainability and resilience of agricultural systems.

Chapter 2: Water Management and Conservation

Water is a vital resource for agriculture, underpinning food production and ecosystem health. However, the increasing demand for water, coupled with the impacts of climate change, has led to significant challenges in agricultural water management. This chapter examines these challenges, focusing on the overuse and depletion of water resources and the effects of climate change on water availability. It then explores nature-based solutions, including agroforestry systems, wetland restoration, rainwater harvesting, and constructed wetlands, which offer sustainable strategies for improving water management and conservation in agricultural landscapes.

Challenges in Agricultural Water Use

Water management is one of the most pressing issues in modern agriculture. Effective and sustainable water use is critical for crop production, livestock management, and overall ecosystem health. However, agriculture faces significant challenges related to water overuse, depletion, and the impacts of climate change. Understanding these challenges is essential for developing sustainable water management practices that ensure the long-term viability of agricultural systems.

Overuse and Depletion of Water Resources

Overuse and depletion of water resources are major concerns in agriculture, primarily driven by the increasing demand for food, fiber, and bioenergy. Intensive agricultural practices often require large amounts of water, leading to the over-extraction of surface and groundwater resources.

Groundwater Overdraft

Groundwater, a critical source of irrigation water, is being extracted at unsustainable rates in many agricultural regions. Overdrafting occurs when water is pumped from aquifers faster than it can be naturally replenished, leading to declining water tables. This depletion can result in reduced water availability for future agricultural use, increased pumping costs, and land subsidence, which can damage infrastructure and reduce soil productivity.

Surface Water Diversions

The diversion of rivers and streams for irrigation purposes can significantly alter natural water flows, affecting aquatic ecosystems and downstream water users. Excessive water withdrawals can reduce river flow, degrade water quality, and harm fish and wildlife habitats. In some cases, rivers have even dried up before reaching their natural endpoints, causing severe ecological and economic impacts.

Inefficient Irrigation Practices

Traditional irrigation methods, such as flood and furrow irrigation, are often inefficient, leading to significant water losses through evaporation, runoff, and seepage. These methods can also contribute to soil salinization, reducing soil fertility and crop yields. Despite advances in irrigation technology, many farmers continue to use inefficient practices due to cost, lack of awareness, or limited access to modern irrigation systems.

Water Pollution

Agricultural activities can pollute water sources through runoff containing fertilizers, pesticides, and sediments. This pollution not only affects water quality but also increases the demand for clean water for irrigation and other uses. Contaminated water sources can harm aquatic ecosystems, reduce biodiversity, and pose health risks to humans and livestock.

Addressing the overuse and depletion of water resources requires a combination of policy measures, technological innovations, and changes in agricultural practices. Sustainable water management strategies, such as improving irrigation efficiency, adopting water-saving technologies, and implementing integrated water resource management, are essential for ensuring the long-term sustainability of water resources in agriculture.

Impact of Climate Change on Water Availability

Climate change is profoundly affecting water availability and distribution, posing significant challenges to agricultural water management. Changes in temperature, precipitation patterns, and the frequency and intensity of extreme weather events are altering the hydrological cycle and impacting water resources.

Changes in Precipitation Patterns

Climate change is causing shifts in precipitation patterns, with some regions experiencing more intense rainfall and others facing prolonged droughts. These changes can lead to irregular water availability, making it difficult for farmers to plan and manage irrigation schedules. In areas prone to heavy rainfall, increased runoff can lead to soil erosion and nutrient loss, while prolonged droughts can stress water supplies and reduce crop yields.

Increased Frequency and Intensity of Extreme Weather Events

Extreme weather events, such as storms, floods, and droughts, are becoming more frequent and severe due to climate change. These events can cause immediate and long-term damage to agricultural systems, including crop loss, soil erosion, and infrastructure damage. Droughts can deplete water sources, while floods can contaminate water supplies and destroy crops.

Rising Temperatures and Evapotranspiration

Higher temperatures increase evapotranspiration rates, leading to greater water loss from soil and plants. This heightened water demand can stress existing water supplies, particularly in arid and semi-arid regions. Warmer temperatures can also reduce snowpack and alter snowmelt timing, affecting the availability of water for irrigation during critical growing periods.

Glacial Melt and Sea Level Rise

In regions dependent on glacial meltwater for irrigation, such as parts of South Asia and South America, the accelerated melting of glaciers due to climate change poses a significant threat to water availability. Additionally, sea level rise can lead to saltwater intrusion into freshwater aquifers and coastal agricultural lands, reducing the availability of fresh water for irrigation and harming crop productivity.

Water Quality Degradation

Climate change can also affect water quality by altering the temperature and flow patterns of rivers and lakes. Warmer water temperatures can increase the prevalence of harmful algal blooms and waterborne pathogens, while changes in flow patterns can concentrate pollutants and reduce water quality. Poor water quality further complicates water management in agriculture, as clean water becomes more scarce and expensive.

To address the impacts of climate change on water availability, it is crucial to adopt adaptive water management practices. These include implementing efficient irrigation systems, developing drought-resistant crop varieties, improving water storage infrastructure, and enhancing watershed management. Additionally, integrating climate change projections into water resource planning and decision-making can help build resilient agricultural systems capable of withstanding the challenges posed by a changing climate.

In conclusion, the challenges in agricultural water use, including overuse, depletion, and the impacts of climate change, require comprehensive and sustainable management strategies. By understanding these challenges and adopting nature-based and technological solutions, agriculture can move towards more sustainable and resilient water management practices.

Nature-Based Water Management Solutions

Nature-based water management solutions provide innovative and sustainable strategies to address the pressing challenges of water use and conservation in agriculture. These solutions leverage natural processes and ecosystems to enhance water availability, improve water quality, and increase the resilience of agricultural systems to climate change. By integrating practices such as agroforestry, wetland restoration, rainwater harvesting, and constructed wetlands, farmers can sustainably manage water resources, protect ecosystems, and support long-term agricultural productivity.

Agroforestry Systems

Agroforestry systems integrate trees and shrubs into agricultural landscapes, creating a synergistic relationship between crops, livestock, and woody perennials. This practice offers multiple benefits, including improved water management.

Enhanced Water Infiltration and Reduced Runoff

Trees and shrubs in agroforestry systems enhance soil structure with their extensive root systems, promoting water infiltration and reducing surface runoff. This helps recharge groundwater, maintain soil moisture, and reduce soil erosion. By capturing and storing rainfall, agroforestry systems contribute to a more stable water supply for crops and livestock.

Improved Microclimate and Soil Moisture

The presence of trees and shrubs can modify the microclimate in agricultural fields by providing shade and reducing wind speed. This helps to lower soil and air temperatures, reducing evapotranspiration rates and conserving soil moisture. As a result, crops and pastures benefit from a more favorable growing environment, especially during dry periods.

Increased Biodiversity and Ecosystem Services

Agroforestry systems promote biodiversity by providing habitats for various plant and animal species. This biodiversity supports ecosystem services, such as pollination and pest control, which are vital for crop productivity. Additionally, the diverse plant species in agroforestry systems contribute to nutrient cycling and soil fertility, enhancing overall agricultural sustainability.

Carbon Sequestration and Climate Resilience

Trees and shrubs in agroforestry systems sequester carbon, helping to mitigate climate change. Their presence also increases the resilience of agricultural landscapes to climate variability and extreme weather events. By stabilizing soil and regulating water cycles, agroforestry systems reduce the vulnerability of crops and livestock to droughts, floods, and storms.

Economic and Social Benefits

Agroforestry systems can diversify farm income by providing additional products, such as fruits, nuts, timber, and fodder. This diversification reduces the economic risks associated with monocropping and enhances the livelihoods of farming communities. Furthermore, agroforestry practices often require less input in terms of water, fertilizers, and pesticides, reducing production costs and promoting sustainable land management.

In conclusion, agroforestry systems offer a holistic approach to water management in agriculture. By integrating trees and shrubs into farming practices, these systems enhance water infiltration, improve soil moisture, support biodiversity, sequester carbon, and provide economic benefits, contributing to the overall sustainability and resilience of agricultural landscapes.

Wetland Restoration

Wetland restoration involves rehabilitating degraded wetlands to restore their natural functions and ecosystem services. Wetlands play a crucial role in water management, providing numerous benefits for agricultural landscapes.

Natural Water Filtration and Quality Improvement

Wetlands act as natural filters, trapping sediments, nutrients, and pollutants from surface runoff before they reach water bodies. This improves water quality, benefiting both agricultural use and downstream ecosystems. Restored wetlands help reduce the load of fertilizers and pesticides entering rivers and lakes, mitigating the impacts of agricultural runoff on aquatic habitats.

Flood Mitigation and Water Storage

Wetlands can absorb and store large volumes of water, acting as natural buffers during heavy rainfall and flooding events. By temporarily holding excess water, wetlands reduce the intensity and frequency of downstream flooding, protecting agricultural fields and infrastructure. This water storage capacity also helps maintain groundwater levels and provides a steady supply of water during dry periods.

Biodiversity Enhancement and Habitat Provision

Wetlands support a rich diversity of plant and animal species, many of which are adapted to wetland conditions. Restoring wetlands creates habitats for various species, including birds, amphibians, fish, and invertebrates. This biodiversity is essential for ecosystem health and resilience, contributing to pest control, pollination, and nutrient cycling in adjacent agricultural areas.

Carbon Sequestration and Climate Regulation

Wetlands are significant carbon sinks, storing carbon in their vegetation and soils. Restoring wetlands enhances their ability to sequester carbon, contributing to climate change mitigation. Additionally, wetlands regulate local climates by maintaining humidity and moderating temperatures, creating a more stable environment for agriculture.

Socio-Economic and Cultural Benefits

Restored wetlands can provide socio-economic benefits, including opportunities for eco-tourism, recreation, and education. They also hold cultural and historical significance for many communities, offering spaces for traditional practices and community activities. By enhancing the multifunctionality of landscapes, wetland restoration supports the well-being of rural and urban populations.

In summary, wetland restoration is a powerful nature-based solution for water management in agriculture. By improving water quality, mitigating floods, enhancing biodiversity, sequestering carbon, and providing socio-economic benefits, restored wetlands contribute to the sustainability and resilience of agricultural systems and the broader environment.

Rainwater Harvesting

Rainwater harvesting is the collection and storage of rainwater for agricultural, domestic, or industrial use. This practice is a sustainable solution to water scarcity, providing an alternative water source and reducing the dependency on groundwater and surface water supplies.

Water Supply Augmentation

Rainwater harvesting systems capture and store rainwater from rooftops, paved surfaces, and other catchment areas. This stored water can be used for irrigation, livestock, and other agricultural needs, supplementing traditional water sources. By capturing rainwater during wet seasons, farmers can ensure a more reliable water supply during dry periods, reducing the risk of crop failure and supporting continuous agricultural production.

Reduced Pressure on Groundwater and Surface Water

Harvesting rainwater decreases the reliance on groundwater and surface water sources, which are often over-extracted for agricultural use. This reduction in demand helps prevent the depletion of aquifers and protects river and stream ecosystems from excessive water withdrawals. By balancing the use of different water sources, rainwater harvesting promotes sustainable water management and conserves vital water resources.

Flood Mitigation and Soil Erosion Control

Rainwater harvesting systems can help manage stormwater runoff, reducing the risk of flooding and soil erosion. By capturing and storing rainwater, these systems prevent large volumes of water from flowing rapidly over land, which can cause soil erosion and damage to crops and infrastructure. This controlled release of stored rainwater can be used strategically to irrigate fields, enhancing soil moisture and reducing erosion.

Improved Water Quality and Nutrient Management

Rainwater is generally of good quality, free from the pollutants and salts often found in groundwater and surface water. Using rainwater for irrigation can improve soil and crop health by avoiding the build-up of salts and chemicals that can result from poor-quality water sources. Additionally, rainwater harvesting systems can be integrated with filtration and treatment processes to ensure the water meets quality standards for agricultural use.

Cost-Effective and Environmentally Friendly

Rainwater harvesting systems are relatively low-cost and easy to install and maintain. They can be adapted to various scales, from small household systems to large community or farm-based installations. The use of rainwater reduces the need for energy-intensive water pumping and treatment, lowering operational costs and minimizing the environmental footprint of water use in agriculture.

In conclusion, rainwater harvesting is a practical and sustainable approach to water management in agriculture. By augmenting water supply, reducing pressure on existing water sources, mitigating floods, improving water quality, and offering cost-effective solutions, rainwater harvesting supports resilient and sustainable agricultural practices.

Constructed Wetlands

Constructed wetlands are engineered systems designed to mimic the natural processes of wetlands for the treatment of wastewater and stormwater. These systems use vegetation, soil, and microbial activity to remove contaminants and improve water quality, providing an effective and sustainable water management solution for agriculture.

Wastewater Treatment and Recycling

Constructed wetlands are used to treat agricultural runoff, livestock wastewater, and other sources of contaminated water. The vegetation and microorganisms in these systems break down organic matter, absorb nutrients, and filter out pollutants, producing cleaner water that can be reused for irrigation and other agricultural purposes. This recycling of water reduces the demand for freshwater and promotes sustainable water use.

Nutrient Management and Pollution Control

Constructed wetlands are highly effective at removing nutrients such as nitrogen and phosphorus from wastewater. These nutrients, if not managed properly, can cause eutrophication in water bodies, leading to algal blooms and degradation of aquatic ecosystems. By removing excess nutrients, constructed wetlands help protect water quality and reduce the environmental impact of agricultural activities.

Habitat Creation and Biodiversity Enhancement

In addition to their water treatment functions, constructed wetlands provide valuable habitats for a variety of plant and animal species. These systems support biodiversity by creating environments similar to natural wetlands, attracting birds, insects, amphibians, and other wildlife. This enhancement of biodiversity contributes to the overall health and resilience of agricultural landscapes.

Flood Control and Water Storage

Constructed wetlands can temporarily store excess water during heavy rainfall, reducing the risk of flooding in agricultural areas. By managing stormwater runoff and allowing it to percolate slowly into the ground or be released gradually, these systems help mitigate flood damage and protect crops and infrastructure.

Aesthetic and Recreational Benefits

Constructed wetlands can enhance the aesthetic value of agricultural landscapes and provide recreational opportunities for communities. These systems can be designed to blend with the natural environment, creating attractive green spaces that offer educational and recreational benefits.

In summary, constructed wetlands are an innovative and sustainable solution for water management in agriculture. By treating wastewater, managing nutrients, enhancing biodiversity, controlling floods, and providing additional benefits, constructed wetlands contribute to the sustainability and resilience of agricultural systems.

Chapter 3: Biodiversity Enhancement

Biodiversity is a critical component of sustainable agriculture, providing essential ecosystem services that support crop production, soil health, and resilience to environmental changes. This chapter delves into the importance of biodiversity in agricultural systems, highlighting the various ecosystem services it provides. It then explores nature-based strategies for enhancing biodiversity on farms, including the use of hedgerows and buffer strips, the creation of pollinator habitats, the implementation of integrated pest management (IPM), and the adoption of agroecological practices. By understanding and enhancing biodiversity, farmers can improve agricultural productivity, sustainability, and resilience.

Importance of Biodiversity in Agriculture

Biodiversity, or the variety of life in all its forms, plays a fundamental role in agriculture. It encompasses the diversity of plants, animals, microorganisms, and ecosystems, all of which contribute to the health and productivity of agricultural systems. This section explores the importance of biodiversity in agriculture by examining the ecosystem services it provides and its overall impact on sustainability and resilience.

Ecosystem Services Provided by Biodiversity

Ecosystem services are the benefits that humans derive from natural ecosystems. In agricultural systems, these services are critical for maintaining productivity, enhancing resilience, and ensuring environmental sustainability. Biodiversity underpins a wide range of ecosystem services that directly and indirectly support agriculture.

Nutrient Cycling

Biodiversity enhances nutrient cycling by involving various organisms in the decomposition of organic matter and the transformation of nutrients into forms that plants can use. Soil microorganisms, such as bacteria and fungi, break down plant and animal residues, releasing essential nutrients like nitrogen, phosphorus, and potassium. This process not only fertilizes the soil naturally but also improves soil structure and fertility, promoting healthy plant growth.

Pollination

Many crops rely on pollinators, such as bees, butterflies, and other insects, to produce fruits, seeds, and nuts. Biodiverse agricultural landscapes support a rich community of pollinators, which are crucial for the successful reproduction of many plants. Pollination services provided by wild and managed pollinators contribute significantly to crop yields and quality. Without a diverse population of pollinators, the productivity of many crops would decline, threatening food security.

Pest and Disease Control

Biodiversity plays a vital role in controlling agricultural pests and diseases. Predatory insects, birds, and mammals help regulate pest populations by preying on herbivorous insects and rodents that can damage crops. Diverse plant communities can also reduce the spread of diseases by interrupting the life cycles of pathogens and pests. Integrated pest management (IPM) strategies leverage biodiversity to minimize the use of chemical pesticides, reducing environmental contamination and promoting sustainable pest control.

Soil Health and Structure

Biodiversity contributes to soil health by maintaining a balance of soil organisms that decompose organic matter, fix nitrogen, and form symbiotic relationships with plant roots. Earthworms, ants, and other soil-dwelling creatures aerate the soil and improve its structure, enhancing water infiltration and root penetration. Healthy soils with diverse microbial communities are more resilient to erosion, compaction, and degradation, providing a stable foundation for agricultural productivity.

Water Regulation and Purification

Biodiversity influences the regulation and purification of water within agricultural landscapes. Vegetation, including trees, shrubs, and grasses, helps manage water flow by absorbing rainfall, reducing runoff, and promoting groundwater recharge. Wetlands and riparian buffers filter sediments, nutrients, and pollutants from surface water, improving water quality and protecting downstream ecosystems. These natural processes ensure a reliable supply of clean water for irrigation and other agricultural uses.

Climate Regulation and Resilience

Biodiverse agricultural systems are better equipped to adapt to and mitigate the impacts of climate change. Diverse plant species can stabilize microclimates, reduce temperature fluctuations, and protect crops from extreme weather events. Biodiversity also enhances carbon sequestration, with different plant species and soil organisms storing carbon in biomass and soils. This contributes to climate change mitigation by reducing greenhouse gas concentrations in the atmosphere.

Genetic Diversity and Crop Improvement

Genetic diversity within crops and livestock is crucial for breeding programs and the development of new varieties that can withstand pests, diseases, and environmental stresses. Wild relatives of cultivated plants and traditional varieties offer a reservoir of genetic traits that can be used to improve modern crops. Maintaining genetic diversity in agricultural systems ensures a greater capacity for adaptation and resilience in the face of changing conditions.

Cultural and Economic Benefits

Biodiversity provides cultural, aesthetic, and recreational benefits that contribute to the well-being of farming communities. Traditional farming practices often rely on a diverse array of crops and livestock, preserving cultural heritage and knowledge. Biodiverse landscapes can also support eco-tourism and other economic activities that generate additional income for farmers and rural communities.

In conclusion, biodiversity is integral to the functioning and sustainability of agricultural systems. It supports essential ecosystem services, including nutrient cycling, pollination, pest control, soil health, water regulation, climate resilience, and genetic diversity. By fostering biodiversity, farmers can enhance the productivity and resilience of their agricultural systems, ensuring long-term sustainability and food security. Efforts to protect and enhance biodiversity in agriculture are therefore critical for achieving a balance between food production, environmental health, and human well-being.

Nature-Based Strategies for Enhancing Biodiversity

Enhancing biodiversity in agricultural systems is essential for promoting sustainability, resilience, and productivity. Various nature-based strategies can be implemented to achieve this goal, including the use of hedgerows and buffer strips, creating pollinator habitats, adopting integrated pest management (IPM), and practicing agroecology. These approaches leverage natural processes and ecosystems to foster a diverse and healthy agricultural landscape.

Hedgerows and Buffer Strips

Hedgerows and buffer strips are linear features of vegetation planted along field boundaries, waterways, and other landscape elements. These strips of vegetation play a significant role in enhancing biodiversity within agricultural landscapes.

Habitat Provision

Hedgerows and buffer strips provide habitats for a wide range of species, including birds, insects, mammals, and plants. These areas offer food, shelter, and breeding sites, supporting biodiversity and creating ecological corridors that connect different parts of the landscape. By linking habitats, they facilitate the movement of species and genetic exchange, which is crucial for maintaining healthy populations.

Erosion Control and Water Quality

The vegetation in hedgerows and buffer strips helps stabilize soil and reduce erosion. Their root systems bind the soil, preventing it from being washed away by rain or irrigation. Additionally, these vegetated areas filter runoff, trapping sediments, nutrients, and pollutants before they reach water bodies. This improves water quality and protects aquatic ecosystems from contamination.

Pest Control and Pollination

Hedgerows and buffer strips support beneficial insects, such as predators and parasitoids that help control agricultural pests. They also provide habitats for pollinators, such as bees and butterflies, which are essential for the pollination of many crops. By enhancing the presence of these beneficial organisms, hedgerows and buffer strips contribute to natural pest control and pollination services.

Microclimate Regulation

These vegetated strips can modify the microclimate within agricultural fields by providing shade, reducing wind speed, and increasing humidity. This can create more favorable conditions for crop growth and reduce the stress caused by extreme weather events. The presence of hedgerows and buffer strips can also enhance soil moisture retention, benefiting crops during dry periods.

In summary, hedgerows and buffer strips are effective nature-based strategies for enhancing biodiversity in agricultural systems. They provide habitats, control erosion, improve water quality, support pest control and pollination, and regulate the microclimate, contributing to the overall health and sustainability of agricultural landscapes.

Pollinator Habitats

Pollinators, such as bees, butterflies, birds, and bats, are vital for the reproduction of many crops and wild plants. Creating and maintaining habitats for pollinators is crucial for enhancing biodiversity and ensuring the sustainability of agricultural production.

Floral Diversity

Providing a diverse array of flowering plants throughout the growing season is essential for supporting pollinators. Different pollinators have varied preferences for flower shapes, colors, and blooming times. By planting a mix of native wildflowers, shrubs, and trees, farmers can ensure a continuous supply of nectar and pollen, attracting and sustaining a diverse pollinator community.

Habitat Features

Pollinator habitats should include nesting sites, shelter, and water sources. For bees, this means providing bare ground for ground-nesting species and installing bee hotels for cavity-nesting bees. Butterflies and other insects benefit from areas with tall grasses and shrubs for shelter. Providing shallow water sources, such as birdbaths or small ponds with landing sites, is also important for pollinator hydration.

Reduced Pesticide Use

Minimizing the use of pesticides is critical for protecting pollinators. Pesticides, especially insecticides, can be harmful or lethal to pollinators. Integrated pest management (IPM) practices that prioritize biological control, habitat management, and the use of less toxic substances can help reduce pesticide impacts. Additionally, applying pesticides during times when pollinators are less active, such as early morning or late evening, can mitigate risks.

Landscape Connectivity

Creating connected habitats is vital for the movement and survival of pollinators. Establishing pollinator-friendly corridors that link different habitat patches allows pollinators to move freely across the landscape, accessing diverse food sources and nesting sites. This connectivity also helps maintain genetic diversity within pollinator populations.

By creating and maintaining pollinator habitats, farmers can enhance biodiversity and improve crop yields through increased pollination services. Supporting pollinators not only benefits agriculture but also contributes to the conservation of these essential species and the overall health of ecosystems.

Integrated Pest Management (IPM)

Integrated pest management (IPM) is an environmentally friendly approach to pest control that combines biological, cultural, physical, and chemical methods to manage pest populations in a sustainable manner. IPM enhances biodiversity by promoting the use of natural pest control mechanisms.

Biological Control

IPM emphasizes the use of natural predators, parasitoids, and pathogens to control pest populations. Introducing and conserving beneficial organisms, such as ladybugs, lacewings, and predatory wasps, helps keep pest numbers in check. These biological control agents thrive in diverse ecosystems, where they can find food and shelter.

Cultural Practices

Cultural practices, such as crop rotation, intercropping, and selecting pest-resistant crop varieties, can reduce pest pressure. Crop rotation disrupts pest life cycles, while intercropping creates a more complex habitat that can confuse and deter pests. Planting pest-resistant varieties reduces the need for chemical interventions, supporting a more diverse and resilient agricultural system.

Physical and Mechanical Control

Physical and mechanical control methods, such as traps, barriers, and manual removal, are important components of IPM. These methods target specific pests without harming non-target organisms. For example, pheromone traps can attract and capture pests, while row covers can protect crops from insect damage.

Chemical Control

When necessary, IPM incorporates the judicious use of chemical controls, selecting the least toxic and most targeted pesticides available. This minimizes the impact on non-target species, including beneficial insects and pollinators. Applying chemicals in a way that reduces exposure to non-target organisms, such as spot treatments and timing applications to avoid peak pollinator activity, is also crucial.

Monitoring and Decision-Making

Regular monitoring of pest populations and crop health is a cornerstone of IPM. Farmers use tools such as pest traps, field scouting, and remote sensing to assess pest pressure and determine the most appropriate control measures. This data-driven approach ensures that interventions are timely and effective, reducing the reliance on broad-spectrum pesticides.

In conclusion, IPM is a holistic approach to pest management that enhances biodiversity by promoting natural pest control mechanisms and reducing the use of harmful chemicals. By integrating biological, cultural, physical, and chemical methods, IPM supports sustainable agricultural practices and contributes to the health and resilience of agricultural ecosystems.

Agroecology

Agroecology is an ecological approach to agriculture that integrates principles of ecology into agricultural production systems. It emphasizes diversity, sustainability, and the use of natural processes to enhance biodiversity and agricultural productivity.

Diversified Farming Systems

Agroecology promotes diversified farming systems that include a variety of crops, livestock, and trees. This diversity enhances ecosystem resilience and stability, reducing the risk of pest outbreaks and crop failures. Polycultures, agroforestry, and mixed cropping systems are examples of diversified practices that mimic natural ecosystems.

Soil Health and Fertility

Maintaining healthy soils is a key principle of agroecology. Practices such as cover cropping, composting, and reduced tillage enhance soil organic matter, improve soil structure, and support diverse soil microbial communities. Healthy soils are more productive and resilient, supporting diverse plant and animal life.

Resource Efficiency

Agroecology focuses on efficient resource use, minimizing external inputs such as synthetic fertilizers and pesticides. By relying on natural processes, such as biological nitrogen fixation, nutrient recycling, and natural pest control, agroecology reduces environmental impacts and promotes sustainability.

Social and Economic Benefits

Agroecology also emphasizes the social and economic dimensions of agriculture. It supports small-scale farmers and local communities, promotes food sovereignty, and values traditional knowledge and practices. By fostering a closer connection between farmers and their environment, agroecology enhances the cultural and economic resilience of rural areas.

Climate Resilience

Agroecological practices increase the resilience of agricultural systems to climate change. Diverse cropping systems, healthy soils, and integrated landscape management enhance the capacity of farms to withstand extreme weather events and adapt to changing conditions.

In summary, agroecology is a comprehensive approach to farming that integrates ecological principles to enhance biodiversity, sustainability, and resilience. By promoting diversified farming systems, improving soil health, and emphasizing resource efficiency, agroecology supports productive and sustainable agricultural landscapes.

In conclusion, nature-based strategies such as hedgerows and buffer strips, pollinator habitats, integrated pest management, and agroecology are essential for enhancing biodiversity in agricultural systems. These approaches leverage natural processes and ecosystems to create sustainable, resilient, and productive agricultural landscapes.

Chapter 4: Carbon Sequestration and Climate Mitigation

Agriculture plays a significant role in climate change, both as a source of greenhouse gas emissions and as a potential solution through carbon sequestration. This chapter examines the impact of agricultural practices on climate change, focusing on the emissions generated by various farming activities. It then explores nature-based carbon sequestration methods, including afforestation and reforestation, soil carbon sequestration, and perennial cropping systems. By understanding and implementing these strategies, agriculture can contribute to climate mitigation efforts while enhancing soil health and productivity.

Agriculture's Role in Climate Change

Agriculture is both a significant contributor to climate change and a potential part of the solution. Understanding the role of agriculture in climate change involves examining the sources of greenhouse gas (GHG) emissions from agricultural practices and exploring ways to mitigate these emissions. This section delves into the various ways agriculture contributes to GHG emissions and highlights the importance of addressing these emissions to combat climate change.

Greenhouse Gas Emissions from Agriculture

Agriculture contributes to climate change through the emission of several greenhouse gases, including carbon dioxide (CO_2), methane (CH_4), and nitrous oxide (N_2O). These emissions result from a variety of agricultural activities, such as livestock production, rice cultivation, soil management, and the use of synthetic fertilizers and pesticides.

Carbon Dioxide (CO_2) Emissions

CO_2 is released primarily through the burning of fossil fuels in agricultural machinery, deforestation, and soil disturbance. The following activities are key contributors:

- Deforestation and Land-Use Change: Converting forests and natural landscapes into agricultural land releases significant amounts of CO_2 stored in trees and soil. Deforestation for agriculture not only contributes to carbon emissions but also reduces the capacity of ecosystems to sequester carbon, exacerbating climate change.

- Soil Disturbance: Tilling and other soil management practices disrupt soil structure and accelerate the decomposition of organic matter, releasing CO_2 into the atmosphere. Intensive tillage, in particular, exposes large amounts of soil organic carbon to oxidation, contributing to higher CO_2 emissions.

- Fossil Fuel Use: The use of fossil fuels in tractors, irrigation pumps, and other agricultural machinery generates CO_2 emissions. Additionally, the production and transportation of agricultural inputs, such as fertilizers and pesticides, involve significant fossil fuel consumption, further contributing to CO_2 emissions.

Methane (CH_4) Emissions

Methane is a potent greenhouse gas with a global warming potential much higher than that of CO_2. Agriculture is a major source of CH_4 emissions, primarily from livestock production and rice paddies:

- Enteric Fermentation: Ruminant animals, such as cattle, sheep, and goats, produce methane during digestion through a process known as enteric fermentation. Methane is released when these animals burp, making livestock production one of the largest sources of agricultural methane emissions.

- Manure Management: Methane is also produced from the anaerobic decomposition of organic matter in animal manure. The way manure is stored and managed can significantly impact the amount of methane emitted. For example, storing manure in lagoons or pits where it decomposes anaerobically generates more methane than using dry or aerobic methods.

- Rice Cultivation: Flooded rice paddies create anaerobic conditions that promote the production of methane by methanogenic bacteria. Methane is released from the soil to the atmosphere through diffusion and through the rice plants' aerenchyma (specialized tissue that allows gas exchange).

Nitrous Oxide (N_2O) Emissions

Nitrous oxide is another potent greenhouse gas, with a global warming potential even greater than that of methane. It is primarily emitted from soils as a result of microbial processes and the use of nitrogen-based fertilizers:

- Synthetic Fertilizers: The application of synthetic nitrogen fertilizers, such as urea and ammonium nitrate, increases the availability of nitrogen in the soil. Microorganisms in the soil convert this nitrogen into nitrous oxide through a process called nitrification and denitrification. The inefficiency of fertilizer use, where not all applied nitrogen is taken up by crops, exacerbates N_2O emissions.

- Organic Fertilizers and Manure: The use of organic fertilizers, such as compost and animal manure, also contributes to N_2O emissions through similar microbial processes. The rate of nitrous oxide emissions depends on various factors, including soil type, moisture content, temperature, and the amount and type of fertilizer applied.

- Soil Disturbance: Tillage and other soil management practices that increase soil aeration and disrupt soil structure can enhance microbial activity, leading to higher N_2O emissions. Additionally, soil compaction from heavy machinery can create anaerobic pockets in the soil, promoting the production of nitrous oxide.

Indirect Emissions

Agricultural activities also contribute to indirect GHG emissions. For example, the production and transportation of fertilizers and pesticides generate CO_2 emissions. Land-use changes associated with agriculture, such as draining wetlands for cultivation, can release significant amounts of stored carbon and methane. Furthermore, agricultural runoff containing nitrogen can lead to the formation of nitrous oxide in downstream water bodies through microbial processes.

Addressing greenhouse gas emissions from agriculture is crucial for mitigating climate change. Several strategies can help reduce these emissions:

1. Improved Livestock Management:

- Dietary Changes: Modifying livestock diets to include more digestible feed and additives that reduce methane production can lower enteric fermentation emissions.

- Manure Management: Implementing anaerobic digesters to capture methane from manure and convert it into biogas for energy use can significantly reduce methane emissions. Other practices, such as composting and aerating manure, can also minimize methane production.

2. Sustainable Crop Practices:

- Reduced Tillage: Adopting no-till or reduced-till farming practices can minimize soil disturbance and CO_2 emissions while enhancing soil carbon sequestration.

- Efficient Fertilizer Use: Applying fertilizers more efficiently, using precision agriculture techniques, and adopting nitrogen inhibitors can reduce nitrous oxide emissions. Utilizing organic amendments, such as cover crops and compost, can improve soil health and nutrient cycling, reducing the need for synthetic fertilizers.

3. Agroforestry and Reforestation:

- Tree Planting: Integrating trees into agricultural landscapes through agroforestry practices and reforestation can sequester carbon, enhance biodiversity, and provide additional income streams for farmers.

4. Rice Cultivation Management:

- Water Management: Implementing alternate wetting and drying (AWD) techniques in rice paddies can reduce methane emissions by periodically allowing the soil to dry, disrupting anaerobic conditions.

5. Policy and Education:

- Incentives and Regulations: Governments can promote sustainable agricultural practices through incentives, subsidies, and regulations. Educating farmers about climate-smart practices and providing technical support can facilitate the adoption of these methods.

In conclusion, agriculture plays a significant role in climate change through the emission of greenhouse gases. However, by adopting sustainable practices and innovative technologies, the agricultural sector can reduce its emissions and contribute to global climate mitigation efforts.

Nature-Based Carbon Sequestration Methods

Nature-based carbon sequestration methods offer sustainable solutions to mitigate climate change by capturing and storing carbon dioxide (CO_2) from the atmosphere. These methods enhance the capacity of ecosystems to sequester carbon, contributing to both climate mitigation and the improvement of agricultural landscapes.

This section explores three key nature-based carbon sequestration methods: afforestation and reforestation, soil carbon sequestration, and perennial cropping systems.

Afforestation and Reforestation

Afforestation and reforestation are powerful strategies for sequestering carbon by increasing the area of land covered by trees. While afforestation involves planting trees on lands that have not been previously forested, reforestation focuses on restoring tree cover on lands that were once forests but have been cleared or degraded.

Carbon Sequestration Potential

Trees absorb CO_2 from the atmosphere through photosynthesis and store it as biomass in their trunks, branches, leaves, and roots. Forests also contribute to soil carbon sequestration as organic matter from fallen leaves and dead trees decomposes and becomes part of the soil carbon pool. The carbon sequestration potential of afforestation and reforestation depends on factors such as tree species, planting density, and forest management practices. Fast-growing species and mixed-species plantations can sequester significant amounts of carbon in a relatively short time.

Biodiversity and Ecosystem Services

In addition to carbon sequestration, afforestation and reforestation enhance biodiversity by creating habitats for various plant and animal species. Forests support ecosystem services such as water regulation, soil stabilization, and climate moderation. By restoring degraded lands, these practices can reverse soil erosion, improve water quality, and provide other environmental benefits.

Economic and Social Benefits

Afforestation and reforestation projects can generate economic benefits for local communities through the sale of timber, non-timber forest products, and carbon credits. These activities can also create employment opportunities in tree planting, forest management, and related industries. Furthermore, forests offer recreational and cultural benefits, enhancing the quality of life for nearby communities.

Challenges and Considerations

Successful afforestation and reforestation require careful planning and management. Selecting appropriate tree species, considering local ecological conditions, and ensuring the long-term maintenance of plantations are crucial for maximizing carbon sequestration and ecosystem benefits. Additionally, these practices should avoid converting valuable natural ecosystems, such as grasslands or wetlands, into forests, as this could lead to unintended ecological consequences.

In summary, afforestation and reforestation are effective nature-based solutions for carbon sequestration. By increasing forest cover, these practices capture and store atmospheric CO_2, enhance biodiversity, and provide multiple ecosystem services and socio-economic benefits. Careful planning and management are essential to ensure the success and sustainability of afforestation and reforestation projects.

Soil Carbon Sequestration

Soil carbon sequestration involves increasing the amount of organic carbon stored in soils through sustainable land management practices. Healthy soils with high organic matter content not only sequester carbon but also support agricultural productivity and ecosystem health.

Carbon Storage Mechanisms

Soils store carbon in organic matter, which includes plant residues, roots, and microbial biomass. The carbon cycle in soils is dynamic, with inputs from plant growth and organic amendments balanced by losses from decomposition and erosion. Sustainable practices that enhance soil organic matter increase the net carbon stored in soils, contributing to climate mitigation.

Agricultural Practices for Soil Carbon Sequestration

Several agricultural practices can enhance soil carbon sequestration:

- Cover Cropping: Planting cover crops during fallow periods protects soil from erosion, adds organic matter, and enhances soil structure. Cover crops such as legumes also fix atmospheric nitrogen, improving soil fertility.

- Reduced Tillage: Minimizing soil disturbance through no-till or reduced-till farming preserves soil structure, reduces erosion, and promotes the accumulation of organic matter on the soil surface.

- Organic Amendments: Adding organic materials such as compost, manure, and biochar enriches soil organic matter and improves soil health. These amendments provide nutrients, enhance microbial activity, and increase soil carbon storage.

- Crop Rotation and Diversification: Rotating different crops and incorporating diverse plant species into farming systems enhances soil organic matter and nutrient cycling. This practice also reduces pest and disease pressures and improves overall soil health.

Benefits Beyond Carbon Sequestration

Enhancing soil carbon sequestration offers multiple co-benefits, including improved soil fertility, water retention, and resilience to drought and extreme weather events. Healthy soils support diverse microbial communities that contribute to nutrient cycling and plant health. Additionally, soil carbon sequestration practices can reduce the need for chemical fertilizers and pesticides, lowering production costs and environmental impacts.

Challenges and Implementation

Implementing soil carbon sequestration practices requires knowledge, resources, and incentives for farmers. Transitioning to sustainable land management practices may involve initial costs and changes in traditional farming methods. Policy support, technical assistance, and financial incentives can facilitate the adoption of soil carbon sequestration practices. Monitoring and verifying soil carbon changes are also essential to ensure the effectiveness of these practices in sequestering carbon.

In conclusion, soil carbon sequestration is a vital strategy for climate mitigation and sustainable agriculture. By enhancing soil organic matter through sustainable practices, farmers can sequester carbon, improve soil health, and achieve multiple environmental and economic benefits. Supportive policies and incentives are crucial for widespread adoption and success.

Perennial Cropping Systems

Perennial cropping systems involve the cultivation of crops that grow and produce yields over multiple years without the need for replanting. These systems have significant potential for carbon sequestration and offer numerous ecological and economic benefits.

Carbon Sequestration Potential

Perennial crops, such as fruit and nut trees, shrubs, and certain grasses, sequester carbon in their biomass and root systems. Unlike annual crops, which are harvested and replanted each year, perennials continue to grow and accumulate carbon in their tissues. The extensive root systems of perennial plants also contribute to soil carbon storage by enhancing soil structure and organic matter content.

Soil Health and Erosion Control

Perennial cropping systems improve soil health by maintaining continuous ground cover, reducing erosion, and enhancing soil structure. The deep and extensive roots of perennial plants stabilize soil, increase water infiltration, and reduce runoff. These systems also support soil microbial communities, which play a crucial role in nutrient cycling and soil carbon sequestration.

Biodiversity and Ecosystem Services

Perennial crops can enhance biodiversity by providing habitats for various plant and animal species. Agroforestry systems, which combine perennials with annual crops or livestock, create diverse and resilient agricultural landscapes. These systems support pollinators, beneficial insects, and wildlife, contributing to ecosystem services such as pest control and pollination.

Economic and Agronomic Benefits

Perennial cropping systems can provide stable and diversified income streams for farmers. Crops such as fruits, nuts, and perennial grains offer marketable products with potentially higher value than annual crops. Additionally, perennials require fewer inputs, such as fertilizers and pesticides, reducing production costs and environmental impacts. The long-term nature of perennial systems also allows for sustainable land management and reduced labor requirements.

Examples of Perennial Crops

Perennial crops include:

- Agroforestry Systems: Integrating trees and shrubs into agricultural landscapes through practices such as alley cropping, silvopasture, and windbreaks enhances carbon sequestration and provides multiple benefits.

- Perennial Grains: Crops like Kernza®, a perennial wheatgrass, offer the potential for sustainable grain production with lower environmental impacts compared to annual grains.

- Fruit and Nut Orchards: Cultivating fruit and nut trees sequesters carbon, supports biodiversity, and provides valuable food products.

In summary, perennial cropping systems are an effective nature-based solution for carbon sequestration and sustainable agriculture. By promoting perennial crops, farmers can enhance carbon storage, improve soil health, support biodiversity, and achieve economic benefits. These systems contribute to resilient and sustainable agricultural landscapes, addressing both climate mitigation and adaptation goals.

In conclusion, nature-based carbon sequestration methods, including afforestation and reforestation, soil carbon sequestration, and perennial cropping systems, offer sustainable solutions for mitigating climate change. These practices enhance carbon storage in forests, soils, and perennial plants, while providing multiple ecological, economic, and social benefits. Implementing and supporting these methods are essential for achieving climate resilience and sustainable agricultural development.

Chapter 5: Sustainable Livestock Management

Livestock management plays a crucial role in sustainable agriculture, influencing environmental health, economic viability, and social well-being. This chapter explores the principles and practices of sustainable livestock management, focusing on methods that enhance animal welfare, reduce environmental impacts, and improve productivity. By adopting sustainable practices such as rotational grazing, silvopasture, and integrated crop-livestock systems, farmers can promote soil health, increase biodiversity, and contribute to climate resilience. This chapter provides an in-depth look at these practices, offering insights into how sustainable livestock management can support a more sustainable and resilient agricultural future.

Challenges in Conventional Livestock Farming

Conventional livestock farming, while critical for meeting global food demands, presents significant environmental and health challenges. These issues stem from intensive farming practices that prioritize high productivity often at the expense of ecological balance and animal welfare. This section delves into the environmental and health issues associated with conventional livestock farming, highlighting the need for more sustainable approaches.

Environmental and Health Issues

Conventional livestock farming presents a range of environmental and health issues that have significant implications for sustainability and public health.

Greenhouse Gas Emissions

Livestock farming is a major contributor to greenhouse gas emissions, which drive climate change. Ruminant animals such as cattle, sheep, and goats produce methane (CH_4) during digestion through enteric fermentation. Methane is a potent greenhouse gas, with a global warming potential much higher than carbon dioxide (CO_2). Additionally, manure management, particularly in large-scale operations, releases methane and nitrous oxide (N_2O), another powerful greenhouse gas. The combined emissions from enteric fermentation, manure management, and feed production make livestock farming a significant source of agricultural greenhouse gases.

Land Degradation and Deforestation

Conventional livestock farming often requires extensive land for grazing and feed crop production. This demand for land can lead to deforestation, particularly in tropical regions where forests are cleared to create pastures or grow feed crops like soybeans. Deforestation results in the loss of biodiversity, disruption of ecosystems, and significant carbon emissions from the destruction of carbon-storing trees and soils. Overgrazing by livestock can also lead to land degradation, including soil erosion, desertification, and the loss of productive land.

Water Usage and Pollution

Livestock farming is water-intensive, requiring substantial amounts of water for drinking, feed crop irrigation, and cleaning facilities. In many regions, this high water demand can strain local water resources, leading to water scarcity and competition with other water users. Furthermore, livestock farming generates significant water pollution through runoff from manure, urine, and feed residues. This runoff often contains nutrients (such as nitrogen and phosphorus), pathogens, and chemicals that can contaminate water bodies, leading to eutrophication, algal blooms, and the degradation of aquatic ecosystems.

Biodiversity Loss

The expansion of livestock farming into natural habitats contributes to biodiversity loss. Deforestation, habitat fragmentation, and land conversion for pasture and feed crops reduce the availability of habitats for wildlife. Additionally, the use of pesticides and fertilizers in feed crop production can harm non-target species, including beneficial insects, birds, and aquatic organisms. The loss of biodiversity undermines ecosystem services, such as pollination and pest control, which are vital for agricultural productivity and ecological stability.

Antibiotic Resistance

The overuse of antibiotics in conventional livestock farming poses a serious public health risk. Antibiotics are often used to promote growth and prevent disease in densely populated and unsanitary farming conditions. This widespread use of antibiotics can lead to the development of antibiotic-resistant bacteria, which can spread to humans through direct contact, the environment, and the consumption of animal products. Antibiotic resistance reduces the effectiveness of antibiotics in treating human and animal diseases, posing a significant challenge to public health.

Animal Welfare Concerns

Intensive livestock farming practices often prioritize productivity over animal welfare. Animals in conventional farming systems may be kept in confined spaces with limited access to outdoor areas, natural behaviors, and social interactions. Practices such as tail docking, beak trimming, and castration without pain relief are common in conventional farming. Poor welfare conditions can lead to stress, injury, and disease, affecting both animal health and productivity.

Nutrient Pollution and Soil Degradation

The concentration of large numbers of animals in confined areas generates vast amounts of manure, which can exceed the land's capacity to absorb and utilize the nutrients effectively. Excessive manure application can lead to nutrient pollution, particularly of nitrogen and phosphorus, which can leach into water bodies or volatilize into the atmosphere, contributing to air and water pollution. Over-application of manure can also degrade soil health, leading to imbalances in soil nutrient levels, soil acidification, and reduced soil fertility.

Economic and Social Impacts

The industrialization of livestock farming has led to the consolidation of farms and the decline of small-scale and family-owned farms. This consolidation can have significant social and economic impacts on rural communities, including loss of livelihoods, reduced economic diversity, and social displacement. The focus on high productivity and efficiency can also marginalize traditional farming practices and local knowledge, undermining cultural heritage and community resilience.

Food Safety and Quality

Conventional livestock farming practices can affect food safety and quality. The use of antibiotics, hormones, and other chemicals in animal production can leave residues in meat, milk, and eggs, posing potential health risks to consumers. Intensive farming conditions can also increase the prevalence of foodborne pathogens, such as Salmonella and E. coli, which can contaminate animal products and pose serious health risks to consumers.

Climate Change Vulnerability

Livestock farming is vulnerable to the impacts of climate change, including increased temperatures, changes in precipitation patterns, and the frequency of extreme weather events. These changes can affect feed availability, water resources, and animal health, posing challenges for the sustainability and resilience of livestock production systems.

In conclusion, conventional livestock farming faces significant environmental and health challenges. The high levels of greenhouse gas emissions, land degradation, water usage, biodiversity loss, and antibiotic resistance associated with these practices underscore the need for more sustainable approaches. Addressing these challenges requires a shift towards practices that prioritize ecological balance, animal welfare, and human health, ensuring the long-term sustainability and resilience of livestock farming systems.

Nature-Based Approaches to Livestock Management

Nature-based approaches to livestock management offer sustainable alternatives to conventional practices, enhancing environmental health, animal welfare, and productivity. These methods leverage natural processes and ecosystems to create resilient and efficient livestock systems. This section explores three key nature-based approaches: rotational grazing, silvopasture, and integrated crop-livestock systems.

Rotational Grazing

Rotational grazing is a livestock management practice where animals are moved between different pasture areas to allow forage plants to recover and regrow. This approach mimics natural grazing patterns and offers several benefits for soil health, forage quality, and animal well-being.

Improved Soil Health

Rotational grazing enhances soil health by preventing overgrazing and allowing grasses and other forage plants to recover. The periodic rest periods for each grazing area help maintain soil structure, reduce erosion, and promote the growth of deep-rooted plants. This leads to increased soil organic matter, improved water infiltration, and greater soil fertility.

Enhanced Forage Quality

By rotating livestock, farmers can manage grazing pressure more effectively, ensuring that forage plants are not overgrazed. This promotes a diverse and resilient plant community with higher nutritional value for the animals. Improved forage quality translates to better animal health and productivity, as livestock have access to a variety of nutrient-rich plants.

Increased Biodiversity

Rotational grazing encourages plant diversity by preventing the dominance of a few species and allowing a range of plants to thrive. This diversity supports a more robust ecosystem, providing habitats for various insects, birds, and other wildlife. Increased biodiversity also contributes to the resilience of the pasture ecosystem, reducing the risk of pest and disease outbreaks.

Reduced Soil Erosion and Runoff

Well-managed rotational grazing systems reduce soil erosion and surface runoff by maintaining ground cover and improving soil structure. The roots of forage plants help stabilize the soil, while the plant cover protects against the impact of raindrops. This reduces sediment and nutrient runoff into water bodies, enhancing water quality and protecting aquatic ecosystems.

Animal Health and Welfare

Rotational grazing provides animals with access to fresh, clean forage and reduces the risk of disease associated with overgrazed, contaminated pastures. It also allows livestock to express natural grazing behaviors, improving their overall well-being. Animals are less likely to suffer from nutritional deficiencies and are better able to withstand environmental stresses.

In conclusion, rotational grazing is a sustainable livestock management practice that benefits soil health, forage quality, biodiversity, and animal welfare. By mimicking natural grazing patterns, farmers can create resilient and productive pasture systems that support long-term agricultural sustainability.

Silvopasture

Silvopasture integrates trees, forage, and livestock into a single system, offering multiple benefits for carbon sequestration, biodiversity, and farm productivity. This approach combines forestry and grazing practices to create a diverse and sustainable agricultural landscape.

Carbon Sequestration

Trees in silvopasture systems sequester carbon in their biomass and roots, helping mitigate climate change. The presence of trees also enhances soil carbon storage by adding organic matter through leaf litter and root turnover. This dual carbon sequestration in both vegetation and soil contributes significantly to reducing greenhouse gas concentrations in the atmosphere.

Biodiversity Enhancement

Silvopasture systems support a diverse range of plant and animal species. The combination of trees, shrubs, and forage plants creates varied habitats that attract birds, insects, and other wildlife. This biodiversity enhances ecosystem resilience and provides valuable ecosystem services such as pollination, pest control, and nutrient cycling.

Improved Soil and Water Quality

Trees in silvopasture systems improve soil structure and water infiltration, reducing erosion and runoff. Their root systems stabilize the soil and enhance its ability to retain moisture, which benefits both forage plants and livestock. Improved soil health leads to better water quality by reducing sediment and nutrient runoff into water bodies.

Enhanced Forage Production

Silvopasture systems can improve forage production by creating a microclimate that moderates temperature extremes and protects against wind and heavy rain. The shade provided by trees reduces heat stress on forage plants, leading to more consistent and higher-quality forage production. This microclimate also benefits livestock by providing shelter and reducing heat stress.

Economic and Social Benefits

Silvopasture diversifies farm income by producing multiple products such as timber, fruits, nuts, and high-quality forage. This diversification reduces economic risks and provides additional revenue streams for farmers. Silvopasture also supports cultural and recreational values by creating aesthetically pleasing landscapes and opportunities for eco-tourism and agroforestry education.

In summary, silvopasture is a sustainable livestock management practice that integrates trees, forage, and livestock to enhance carbon sequestration, biodiversity, soil and water quality, and farm productivity. By creating diverse and resilient agricultural landscapes, silvopasture supports both environmental sustainability and economic viability.

Integrated Crop-Livestock Systems

Integrated crop-livestock systems combine crop production and livestock rearing on the same farm, creating synergistic interactions that enhance productivity, soil health, and resource efficiency. This approach leverages the benefits of both systems to create a more sustainable and resilient agricultural model.

Enhanced Soil Fertility

Livestock manure provides valuable organic matter and nutrients that improve soil fertility and structure. When integrated with crop production, manure can be used to fertilize fields, reducing the need for synthetic fertilizers and enhancing soil health. Crop residues, in turn, can be used as livestock feed, creating a closed-loop system that recycles nutrients within the farm.

Diversified Production and Risk Reduction

Integrated crop-livestock systems diversify farm production by combining multiple enterprises, such as grain crops, forage crops, and livestock. This diversification reduces economic risks associated with market fluctuations and environmental stresses. Farmers can benefit from multiple income streams and more stable overall production.

Improved Resource Efficiency

Combining crop and livestock production enhances resource use efficiency by optimizing the use of land, water, and nutrients. Livestock can graze on crop residues and cover crops, reducing feed costs and maximizing the use of available forage. Integrated systems also improve water use efficiency by capturing and recycling water through different farm enterprises.

Increased Biodiversity and Ecosystem Services

Integrated crop-livestock systems promote biodiversity by supporting diverse plant and animal species. Crop rotations, cover cropping, and diversified plantings enhance soil microbial diversity and pest control. Livestock grazing on diverse pastures supports wildlife habitats and contributes to ecosystem services such as pollination and nutrient cycling.

Climate Resilience

Integrated systems enhance climate resilience by creating more diverse and adaptable farming systems. Crop rotations and cover crops improve soil health and water retention, reducing the impacts of drought and extreme weather events. Livestock provide a buffer against crop failure, offering alternative income and food sources during challenging conditions.

Social and Community Benefits

Integrated crop-livestock systems support rural communities by creating employment opportunities and promoting social cohesion. These systems often involve family labor and local knowledge, fostering a sense of community and cultural heritage. Integrated farming practices also offer educational and recreational opportunities, promoting sustainable agriculture awareness.

In conclusion, integrated crop-livestock systems are a sustainable approach to farming that enhances soil fertility, resource efficiency, biodiversity, and climate resilience. By combining crop production and livestock rearing, farmers can create synergistic interactions that support long-term agricultural sustainability and community well-being.

Chapter 6: Agroforestry Systems

Agroforestry systems, which integrate trees and shrubs with crops and livestock, offer a sustainable and multifunctional approach to agriculture. These systems harness the benefits of biodiversity and natural processes to improve productivity, enhance ecosystem services, and increase resilience to climate change. This chapter explores the various types of agroforestry systems, such as alley cropping, forest farming, and silvopasture, and discusses their benefits, implementation strategies, and potential challenges. By understanding and adopting agroforestry practices, farmers can create more sustainable and resilient agricultural landscapes.

Definition and Types of Agroforestry

Agroforestry, the integration of trees and shrubs with crops and livestock, combines agriculture and forestry to create multifunctional and sustainable farming systems. This approach enhances productivity, biodiversity, and sustainability. This section explores the definition of agroforestry and examines various types of agroforestry systems, including alley cropping, forest farming, and riparian buffers.

Alley Cropping

Alley cropping is an agroforestry practice that involves planting rows of trees or shrubs (the "alleys") alongside agricultural crops. This system offers multiple benefits, including enhanced soil fertility, improved microclimate, and increased biodiversity, making it a sustainable and productive approach to farming.

Design and Implementation

In an alley cropping system, trees or shrubs are planted in parallel rows with wide alleys in between where crops are cultivated. The spacing between the rows and the selection of tree and crop species depend on various factors, such as the local climate, soil conditions, and the specific needs of the farm. Common tree species used in alley cropping include nitrogen-fixing trees like black locust (Robinia pseudoacacia) or leguminous shrubs like leucaena (Leucaena leucocephala), which can improve soil fertility by adding nitrogen.

Soil Fertility and Health

One of the primary benefits of alley cropping is the enhancement of soil fertility. The trees and shrubs in the system contribute organic matter through leaf litter and root exudates, which decompose and enrich the soil with nutrients. Additionally, nitrogen-fixing trees and shrubs can increase soil nitrogen levels, reducing the need for synthetic fertilizers. The root systems of the trees also help prevent soil erosion by stabilizing the soil structure and improving water infiltration.

Microclimate Regulation

Alley cropping can create a favorable microclimate for crops by providing shade and wind protection. The presence of trees reduces the intensity of sunlight, lowering temperatures and reducing heat stress on crops. This shading effect is particularly beneficial in regions with high temperatures, as it can improve crop yields and quality. The trees also act as windbreaks, protecting crops from damaging winds and reducing evaporation rates, which helps conserve soil moisture.

Biodiversity and Pest Control

The integration of trees and shrubs in alley cropping systems promotes biodiversity by providing habitats for various plant and animal species. Increased biodiversity can enhance ecosystem resilience and contribute to natural pest control. For example, the presence of diverse plant species can attract beneficial insects and birds that prey on crop pests, reducing the need for chemical pesticides. The diverse root systems also promote a healthy soil microbiome, which supports overall plant health.

Economic Benefits

Alley cropping can provide additional sources of income for farmers through the production of timber, fruits, nuts, or other tree products. This diversification reduces the economic risks associated with relying on a single crop and can improve the overall profitability of the farm. Additionally, the trees in alley cropping systems can be harvested selectively, providing a steady income stream without disrupting the agricultural component of the system.

In summary, alley cropping is a versatile and sustainable agroforestry practice that enhances soil fertility, regulates the microclimate, promotes biodiversity, and offers economic benefits. By integrating trees and crops, farmers can create more resilient and productive agricultural systems.

Forest Farming

Forest farming is an agroforestry practice that involves cultivating high-value crops under the canopy of an existing forest. This method takes advantage of the natural shade and microclimate provided by the forest, allowing farmers to grow a variety of crops that thrive in low-light conditions. Forest farming promotes biodiversity, enhances ecosystem services, and offers economic opportunities for farmers.

Design and Implementation

Forest farming requires careful planning and management to ensure that the needs of both the forest and the crops are met. Suitable crops for forest farming include medicinal herbs, mushrooms, berries, nuts, and ornamental plants. The selection of crops depends on the specific conditions of the forest, such as light availability, soil type, and moisture levels. Proper spacing and pruning of trees may be necessary to optimize light penetration and create a favorable environment for the understory crops.

Biodiversity and Ecosystem Health

Forest farming enhances biodiversity by preserving the forest canopy and maintaining a diverse understory. This practice supports a wide range of plant and animal species, contributing to the overall health and resilience of the forest ecosystem. The presence of diverse crops and tree species can improve soil structure, enhance nutrient cycling, and promote a healthy soil microbiome. Additionally, forest farming can help protect forests from degradation and deforestation by providing economic incentives for their conservation.

Economic Benefits

Forest farming offers farmers the opportunity to diversify their income by cultivating high-value specialty crops. These crops often command premium prices in the market, providing a lucrative alternative to traditional agricultural products. Forest farming can be particularly beneficial for small-scale and resource-limited farmers, as it requires minimal inputs and capital investment. Additionally, the long-term sustainability of forest farming can ensure a stable and continuous source of income.

Sustainable Land Use

By integrating crop production with forest conservation, forest farming promotes sustainable land use practices. This approach minimizes soil erosion, maintains water quality, and reduces the need for chemical inputs such as fertilizers and pesticides. Forest farming also sequesters carbon, contributing to climate change mitigation. The sustainable management of forest resources ensures that both the forest and the crops can thrive, creating a balanced and productive agricultural system.

In summary, forest farming is a sustainable agroforestry practice that leverages the natural benefits of forests to cultivate high-value crops. By enhancing biodiversity, supporting ecosystem health, and providing economic benefits, forest farming offers a viable and sustainable approach to agriculture.

Riparian Buffers

Riparian buffers are vegetated areas located along the banks of rivers, streams, and other water bodies. These buffers play a critical role in protecting water quality, preventing soil erosion, and enhancing biodiversity in agricultural landscapes. By integrating trees, shrubs, and grasses, riparian buffers provide multiple ecosystem services and contribute to the overall health of the watershed.

Water Quality Improvement

Riparian buffers act as natural filters, trapping sediments, nutrients, and pollutants before they enter water bodies. The vegetation in these buffers absorbs excess nutrients, such as nitrogen and phosphorus, from agricultural runoff, reducing the risk of eutrophication and algal blooms in downstream waters. The root systems of trees and shrubs stabilize the soil, preventing erosion and reducing sedimentation in water bodies. This filtration process enhances water quality and protects aquatic habitats.

Erosion Control and Flood Mitigation

The dense vegetation in riparian buffers helps control soil erosion by stabilizing the stream banks and reducing the velocity of surface runoff. During heavy rainfall events, riparian buffers can absorb and slow down the flow of water, reducing the risk of flooding and minimizing damage to agricultural fields and infrastructure. The root systems of trees and shrubs also enhance soil structure, increasing its ability to retain water and reducing surface runoff.

Biodiversity and Habitat Provision

Riparian buffers provide critical habitats for a wide range of plant and animal species. The diverse vegetation supports wildlife, including birds, mammals, insects, and amphibians, creating a rich and dynamic ecosystem. Riparian buffers also serve as corridors for wildlife movement, connecting fragmented habitats and allowing species to migrate and disperse. This connectivity is essential for maintaining healthy populations and promoting genetic diversity.

Climate Regulation and Carbon Sequestration

Trees and shrubs in riparian buffers sequester carbon, helping to mitigate climate change. The vegetation also regulates local microclimates by providing shade, reducing temperature fluctuations, and increasing humidity. These microclimate benefits can extend to adjacent agricultural fields, improving growing conditions for crops and reducing heat stress on livestock.

Economic and Social Benefits

Riparian buffers can enhance the aesthetic value of agricultural landscapes, creating attractive and multifunctional spaces that benefit both farmers and communities. These buffers can also support recreational activities, such as fishing, birdwatching, and hiking, contributing to the local economy and improving quality of life. Additionally, the ecosystem services provided by riparian buffers, such as improved water quality and flood control, reduce the need for costly engineering solutions and enhance the sustainability of agricultural practices.

Implementation and Management

Establishing and maintaining riparian buffers requires careful planning and management. Farmers should select appropriate native species that are well-suited to the local environment and capable of providing the desired ecosystem services. Regular monitoring and maintenance, such as controlling invasive species and managing vegetation growth, are essential to ensure the long-term effectiveness of riparian buffers.

In conclusion, riparian buffers are a valuable agroforestry practice that enhances water quality, controls erosion, supports biodiversity, and provides economic and social benefits. By integrating these vegetated areas into agricultural landscapes, farmers can protect water resources, improve ecosystem health, and promote sustainable land use practices.

Benefits of Agroforestry

Agroforestry systems offer numerous benefits that enhance the sustainability and productivity of agricultural landscapes. By integrating trees and shrubs with crops and livestock, agroforestry practices can improve biodiversity, soil and water quality, and climate resilience. This section delves into these key benefits, highlighting how agroforestry contributes to a more sustainable and resilient agricultural system.

Enhanced Biodiversity

Agroforestry systems significantly enhance biodiversity by creating diverse habitats that support a wide range of plant and animal species. The integration of trees, shrubs, and crops fosters a more complex and varied landscape, promoting ecological interactions and increasing the overall biodiversity of the area.

Habitat Diversity

The presence of trees and shrubs in agricultural landscapes provides habitats for various species, including birds, insects, mammals, and microorganisms. These habitats support nesting, feeding, and breeding activities, contributing to the conservation of wildlife. The diversity of plant species in agroforestry systems also attracts pollinators and beneficial insects, which play crucial roles in maintaining healthy ecosystems and improving crop yields.

Ecological Interactions

Agroforestry systems promote beneficial ecological interactions, such as mutualism and predation, that enhance ecosystem stability. For instance, trees can provide nectar and pollen for pollinators, while also offering shelter for predatory insects that control pest populations. These interactions reduce the need for chemical pesticides and fertilizers, fostering a more balanced and sustainable agricultural environment.

Genetic Diversity

Incorporating a variety of tree and shrub species into farming systems increases genetic diversity, which enhances the resilience of the ecosystem to environmental changes and stresses. This diversity can buffer against the impacts of pests, diseases, and climate extremes, ensuring the long-term sustainability of agricultural production. Genetic diversity also provides a pool of traits that can be used in breeding programs to develop new crop varieties with improved resistance and adaptability.

Landscape Connectivity

Agroforestry systems create corridors and stepping stones that connect fragmented habitats, facilitating the movement and dispersal of species. This connectivity is vital for maintaining genetic diversity and ensuring the survival of wildlife populations. Connected landscapes also support ecosystem functions, such as pollination and seed dispersal, that are essential for maintaining healthy and productive agricultural systems.

In summary, agroforestry enhances biodiversity by providing diverse habitats, promoting ecological interactions, increasing genetic diversity, and improving landscape connectivity. These benefits contribute to the overall health and resilience of agricultural ecosystems, supporting sustainable and productive farming practices.

Improved Soil and Water Quality

Agroforestry practices play a crucial role in improving soil and water quality, which are essential for sustainable agricultural production. By integrating trees and shrubs into farming systems, agroforestry enhances soil structure, fertility, and water retention, while also protecting water resources from pollution and erosion.

Soil Structure and Fertility

The root systems of trees and shrubs in agroforestry systems improve soil structure by creating channels that enhance water infiltration and root penetration. These root systems also help to stabilize the soil, reducing the risk of erosion. The addition of organic matter from leaf litter and root turnover further enhances soil fertility by increasing the availability of nutrients and supporting beneficial microbial activity. This organic matter also improves soil water-holding capacity, reducing the need for irrigation and increasing resilience to drought.

Nutrient Cycling

Trees and shrubs in agroforestry systems contribute to nutrient cycling by capturing nutrients from deep soil layers and making them available to shallow-rooted crops. Nitrogen-fixing trees and shrubs, such as legumes, can enhance soil nitrogen levels, reducing the need for synthetic fertilizers. The presence of diverse plant species in agroforestry systems also promotes a more balanced nutrient distribution, preventing nutrient depletion and soil degradation.

Erosion Control

Agroforestry systems help to prevent soil erosion by protecting the soil surface with vegetation cover and stabilizing the soil with root systems. Trees and shrubs reduce the impact of raindrops on the soil, minimizing surface runoff and soil displacement. This protection is particularly important on sloped or degraded lands, where erosion can significantly reduce soil productivity and contribute to water pollution. By reducing erosion, agroforestry practices help to maintain soil health and agricultural productivity.

Water Quality Protection

Agroforestry systems enhance water quality by filtering and trapping pollutants before they reach water bodies. The vegetation in these systems acts as a buffer, capturing sediments, nutrients, and pesticides from agricultural runoff. This filtration process reduces the risk of water pollution, protecting aquatic ecosystems and ensuring a clean water supply for agricultural and domestic use. Riparian buffers, which are strips of vegetation planted along water bodies, are particularly effective in improving water quality and preventing erosion.

Microclimate Regulation

The presence of trees and shrubs in agroforestry systems can moderate microclimates by providing shade, reducing wind speed, and increasing humidity. These microclimatic effects create a more favorable environment for crop growth and reduce the stress caused by extreme temperatures. Improved microclimates also enhance soil moisture retention, further supporting crop productivity and resilience.

In conclusion, agroforestry improves soil and water quality by enhancing soil structure and fertility, promoting nutrient cycling, controlling erosion, protecting water quality, and regulating microclimates. These benefits contribute to the sustainability and productivity of agricultural systems, ensuring the long-term health of the environment and farming communities.

Climate Resilience

Agroforestry systems enhance climate resilience by increasing the ability of agricultural landscapes to withstand and adapt to the impacts of climate change. By integrating trees and shrubs with crops and livestock, agroforestry practices improve the overall stability and sustainability of farming systems in the face of climate variability and extremes.

Carbon Sequestration

Trees and shrubs in agroforestry systems sequester carbon in their biomass and soils, helping to mitigate climate change by reducing greenhouse gas concentrations in the atmosphere. This carbon storage not only contributes to global climate goals but also enhances soil health and fertility, supporting sustainable agricultural production. Agroforestry systems can significantly increase the carbon sequestration potential of agricultural landscapes compared to conventional farming practices.

Temperature Regulation

The presence of trees and shrubs in agroforestry systems helps to moderate temperatures by providing shade and reducing heat stress on crops and livestock. This shading effect can lower temperatures in the immediate vicinity, creating a more favorable microclimate for agricultural production. Reduced heat stress improves crop yields and quality, and enhances animal welfare and productivity. By buffering against temperature extremes, agroforestry systems increase the resilience of farming systems to heatwaves and other climate-related stresses.

Water Management

Agroforestry practices enhance water management by improving soil water-holding capacity and reducing surface runoff. The deep root systems of trees and shrubs increase water infiltration and groundwater recharge, making water more available to crops during dry periods. This improved water retention reduces the need for irrigation and increases the resilience of agricultural systems to drought. Agroforestry also helps to protect water resources by reducing erosion and sedimentation, ensuring a sustainable water supply for agriculture and other uses.

Biodiversity and Ecosystem Services

The increased biodiversity in agroforestry systems enhances ecosystem resilience by supporting a wide range of species and ecological interactions. Diverse ecosystems are better able to recover from disturbances and adapt to changing conditions. Agroforestry practices promote ecosystem services such as pollination, pest control, and nutrient cycling, which are essential for maintaining productive and resilient agricultural systems. These services reduce the reliance on external inputs and increase the sustainability of farming practices.

Risk Reduction

By diversifying production through the integration of trees, crops, and livestock, agroforestry systems reduce the economic risks associated with market fluctuations and environmental stresses. Farmers can benefit from multiple income streams and greater overall stability, making their operations more resilient to climate-related challenges. This diversification also supports community resilience by providing food security and sustainable livelihoods.

In summary, agroforestry enhances climate resilience by sequestering carbon, regulating temperatures, improving water management, supporting biodiversity, and reducing economic risks. These benefits contribute to the sustainability and stability of agricultural systems, ensuring their ability to adapt and thrive in the face of climate change.

Chapter 7: Urban Agriculture and NbS

Urban agriculture, the practice of cultivating, processing, and distributing food in or around urban areas, offers innovative solutions for enhancing food security, sustainability, and community well-being. By integrating nature-based solutions (NbS) into urban farming, cities can address environmental challenges, promote biodiversity, and improve the quality of urban life. This chapter explores the role of urban agriculture in fostering sustainable urban environments, examining various NbS practices such as rooftop gardens, vertical farming, and community gardens. Through these approaches, urban agriculture can contribute to resilient, green cities and provide fresh, local food for urban residents.

The Rise of Urban Agriculture

Urban agriculture, the practice of cultivating, processing, and distributing food in or around urban areas, has gained significant momentum in recent years. This resurgence is driven by various factors, including the growing awareness of food security issues, the desire for fresh and local produce, and the need to make urban environments more sustainable. As cities expand and populations grow, urban agriculture offers a promising solution to many of the challenges faced by urban communities.

Benefits and Challenges

Urban agriculture offers numerous benefits that extend beyond food production. These benefits include environmental, social, and economic advantages that contribute to the overall well-being of urban communities. However, urban agriculture also faces several challenges that need to be addressed to maximize its potential.

Benefits:

1. Food Security and Nutrition:

Urban agriculture enhances food security by providing a local source of fresh produce, reducing the dependence on external food supplies. It also improves access to nutritious foods, particularly in food deserts where fresh produce is scarce. This can have significant health benefits, reducing the prevalence of diet-related diseases and improving overall public health.

2. Environmental Benefits:

Urban agriculture helps mitigate some of the environmental impacts of urbanization. Green spaces created by urban farms and gardens improve air quality by absorbing pollutants and producing oxygen. They also reduce the urban heat island effect by providing shade and cooling the surrounding areas. Additionally, urban agriculture promotes biodiversity by creating habitats for various plant and animal species.

3. Community Building:

Urban agriculture fosters community engagement and social cohesion. Community gardens and urban farms bring people together, creating spaces for social interaction, education, and collaboration. These projects often involve community members in planning, planting, and harvesting, promoting a sense of ownership and responsibility. They can also provide educational opportunities, teaching residents about sustainable farming practices and healthy eating.

4. Economic Opportunities:

Urban agriculture can stimulate local economies by creating jobs and generating income. It offers opportunities for small-scale entrepreneurs to sell fresh produce, value-added products, and gardening supplies. Additionally, urban farms and gardens can attract tourism and increase property values, contributing to economic development in urban areas.

Challenges:

1. Space and Land Use:

One of the primary challenges of urban agriculture is finding suitable space for farming. Urban areas are densely populated, and land is often scarce and expensive. Securing land for urban agriculture can be difficult, especially in cities with high real estate values. Innovative solutions, such as rooftop gardens and vertical farming, can help address this challenge, but they also require significant investment and technical expertise.

2. Soil and Water Quality:

Urban soils can be contaminated with heavy metals, pollutants, and other harmful substances, posing risks to food safety and human health. Ensuring the quality of soil and water used in urban agriculture is essential but can be challenging. Soil testing, remediation, and the use of raised beds or container gardening can mitigate some of these risks, but they add to the cost and complexity of urban farming.

3. Regulatory and Policy Barriers:

Urban agriculture often faces regulatory and policy barriers, including zoning laws, land use regulations, and health and safety codes. Navigating these regulations can be challenging for urban farmers, especially those without legal or administrative expertise. Advocacy and policy reform are needed to create a supportive regulatory environment that promotes and facilitates urban agriculture.

4. Economic Viability:

Ensuring the economic viability of urban agriculture projects can be challenging. Urban farms and gardens often operate on thin profit margins and may struggle to compete with conventional agricultural products. Access to funding, technical assistance, and market opportunities are critical for the success of urban agriculture initiatives. Building strong business models and forming partnerships with local businesses, governments, and non-profits can help enhance the economic sustainability of urban agriculture.

In conclusion, urban agriculture offers significant benefits for urban communities, including enhanced food security, environmental sustainability, community building, and economic opportunities. However, addressing the challenges of space, soil and water quality, regulatory barriers, and economic viability is essential to unlock its full potential. By overcoming these challenges, urban agriculture can play a vital role in creating resilient, sustainable, and vibrant urban environments.

Nature-Based Solutions in Urban Agriculture

Nature-based solutions (NbS) in urban agriculture integrate natural processes and green infrastructure to enhance the sustainability and resilience of urban food systems. These solutions not only contribute to food security and environmental health but also improve the quality of urban life by creating green spaces, supporting biodiversity, and mitigating climate change. This section explores various NbS practices in urban agriculture, including rooftop gardens, vertical farming, and community gardens, highlighting their benefits, implementation strategies, and potential challenges.

Rooftop Gardens

Rooftop gardens are an innovative approach to urban agriculture that transform the often underutilized spaces atop buildings into productive green areas. These gardens provide numerous environmental, social, and economic benefits, making them a valuable component of sustainable urban development. By growing vegetables, herbs, and flowers on rooftops, cities can enhance food security, improve air quality, and create green spaces that contribute to the well-being of urban residents.

Design and Implementation

Establishing a rooftop garden requires careful planning and design to ensure structural integrity, plant health, and efficient water use. Key considerations include:

- Structural Assessment: The first step is to assess the building's structural capacity to support the additional weight of soil, plants, and water. Engineers and architects must evaluate the roof's load-bearing capacity and make necessary reinforcements.

- Waterproofing and Drainage: Proper waterproofing is essential to protect the building from water damage. Installing a robust drainage system prevents water accumulation and ensures excess water is efficiently removed.

- Growing Medium: Traditional soil may be too heavy for rooftops, so lightweight growing mediums such as a mixture of compost, perlite, and vermiculite are often used. These mediums provide good drainage, aeration, and nutrient retention.

- Irrigation System: Rooftop gardens require an efficient irrigation system, such as drip irrigation or automated watering systems, to ensure plants receive adequate moisture without overloading the roof with water weight.

- Plant Selection: Selecting appropriate plants is crucial for the success of a rooftop garden. Hardy, drought-tolerant, and shallow-rooted plants are ideal. Common choices include herbs, leafy greens, tomatoes, peppers, and flowers.

Environmental Benefits

Rooftop gardens offer several environmental benefits that contribute to urban sustainability:

- Air Quality Improvement: Plants in rooftop gardens absorb carbon dioxide and other pollutants, releasing oxygen and improving air quality. They also capture particulate matter, reducing pollution levels.

- Temperature Regulation: Rooftop gardens help mitigate the urban heat island effect by providing shade and cooling through evapotranspiration. This reduces the need for air conditioning in buildings, lowering energy consumption and greenhouse gas emissions.

- Stormwater Management: Green roofs absorb rainwater, reducing runoff and alleviating pressure on urban drainage systems. This helps prevent flooding and reduces the risk of water pollution from stormwater runoff.

- Biodiversity Enhancement: Rooftop gardens create habitats for birds, insects, and other wildlife, increasing urban biodiversity. They can also serve as stopover sites for migratory species, contributing to regional ecological networks.

Social and Community Benefits

Rooftop gardens provide social and community benefits by creating green spaces in densely populated urban areas:

- Access to Fresh Produce: Rooftop gardens can supply fresh, locally grown produce to urban residents, enhancing food security and promoting healthy eating. This is particularly beneficial in food deserts where access to fresh fruits and vegetables is limited.

- Recreational Spaces: Rooftop gardens offer recreational and therapeutic spaces for residents to relax, socialize, and engage in gardening activities. These green spaces contribute to mental well-being and stress reduction.

- Educational Opportunities: Rooftop gardens serve as living laboratories where people of all ages can learn about sustainable agriculture, horticulture, and environmental stewardship. Schools, community groups, and organizations can use these spaces for educational programs and workshops.

- Community Engagement: Rooftop gardens foster a sense of community by involving residents in the planning, planting, and maintenance of the garden. This collaborative effort strengthens social ties and encourages civic participation.

Economic Benefits

In addition to environmental and social benefits, rooftop gardens can also provide economic advantages:

- Energy Savings: By insulating buildings and reducing the need for air conditioning, rooftop gardens lower energy costs. This can lead to significant savings for building owners and tenants.

- Property Value Increase: Green roofs can enhance the aesthetic appeal of buildings, increasing property values and attracting potential buyers or renters.

- Local Economic Development: Rooftop gardens can create job opportunities in urban farming, horticulture, and landscape maintenance. They can also stimulate local economies by providing fresh produce to nearby markets and restaurants.

Challenges and Considerations

While rooftop gardens offer many benefits, there are also challenges to consider:

- Initial Costs: The installation of rooftop gardens can be expensive, including costs for structural assessments, waterproofing, and irrigation systems. However, the long-term benefits often outweigh the initial investment.

- Maintenance Requirements: Rooftop gardens require ongoing maintenance, including watering, fertilizing, and pest management. Building owners and residents must be committed to regular upkeep to ensure the garden's success.

- Climate and Weather: Rooftop gardens are exposed to harsh weather conditions, including strong winds, intense sunlight, and temperature fluctuations. Selecting resilient plants and providing windbreaks or shading structures can mitigate these challenges.

- Regulatory Barriers: Local building codes and zoning regulations may restrict the installation of rooftop gardens. Navigating these regulations and obtaining necessary permits can be complex.

In conclusion, rooftop gardens are a valuable component of urban agriculture, offering numerous environmental, social, and economic benefits. By transforming unused rooftop spaces into productive green areas, cities can enhance food security, improve air quality, and create vibrant, sustainable communities. Despite the challenges, the potential rewards make rooftop gardens an attractive and viable option for urban development.

Vertical Farming

Vertical farming is an innovative agricultural practice that involves growing crops in vertically stacked layers or integrated structures, often in controlled indoor environments. This method leverages cutting-edge technology and efficient use of space to produce food sustainably in urban settings, addressing challenges such as land scarcity, food security, and environmental impact. Vertical farming offers a promising solution for urban agriculture, providing fresh produce year-round while minimizing resource use.

Design and Implementation

Vertical farming systems can be implemented in various structures, including repurposed warehouses, skyscrapers, shipping containers, and modular units. Key components of vertical farming include:

- Vertical Stacking: Crops are grown in stacked layers, maximizing the use of available space. This can be achieved through shelves, towers, or other vertical arrangements that increase the production area within a limited footprint.

- Hydroponics, Aeroponics, and Aquaponics: Vertical farming often employs soilless cultivation methods. Hydroponics involves growing plants in nutrient-rich water, aeroponics uses mist to deliver nutrients to the roots, and aquaponics combines fish farming with hydroponics, using fish waste to fertilize the plants.

- Controlled Environment Agriculture (CEA): Vertical farms use advanced climate control systems to regulate temperature, humidity, light, and CO_2 levels. This allows for optimal growing conditions and consistent crop production year-round, regardless of external weather conditions.

- LED Lighting: Energy-efficient LED lights provide the necessary spectrum and intensity of light for photosynthesis. These lights can be adjusted to mimic natural sunlight, promoting healthy plant growth and maximizing yields.

Environmental Benefits

Vertical farming offers several environmental advantages that contribute to sustainability:

- Water Efficiency: Vertical farms use significantly less water than traditional farming methods. Recirculating water systems in hydroponics and aeroponics reduce water waste, and controlled environments minimize evaporation. This high water efficiency is particularly beneficial in urban areas facing water scarcity.

- Reduced Land Use: By growing crops vertically, these farms require much less land compared to conventional agriculture. This reduction in land use helps preserve natural ecosystems and reduces the pressure to convert forests and other habitats into farmland.

- Minimized Pesticide Use: Controlled environments in vertical farms are less susceptible to pests and diseases, reducing the need for chemical pesticides. This results in cleaner produce and less environmental pollution.

- Lower Carbon Footprint: Vertical farming reduces the need for transportation by bringing food production closer to urban consumers. This local production minimizes the carbon emissions associated with transporting food over long distances.

Economic Benefits

Vertical farming presents several economic opportunities and benefits:

- Year-Round Production: The ability to control growing conditions enables vertical farms to produce crops year-round, ensuring a continuous supply of fresh produce. This consistent production can lead to stable revenues and reduce the seasonal fluctuations common in traditional agriculture.

- Urban Job Creation: Vertical farms create employment opportunities in urban areas, including roles in farming, maintenance, technology, and distribution. This can stimulate local economies and provide job opportunities for city residents.

- Local Food Systems: By producing food close to consumers, vertical farming supports local food systems and reduces dependence on distant agricultural regions. This can enhance food security and resilience in urban areas.

Social and Community Benefits

Vertical farming can contribute to social well-being and community development:

- Access to Fresh Produce: Urban residents, particularly those in food deserts, gain access to fresh, locally grown produce through vertical farming. This can improve nutrition and health outcomes for urban populations.

- Educational Opportunities: Vertical farms can serve as educational hubs, teaching students and community members about sustainable agriculture, nutrition, and technology. These educational initiatives can raise awareness and foster a culture of sustainability.

- Community Engagement: Vertical farming projects often engage local communities, fostering a sense of connection and shared purpose. Community involvement in urban farming can strengthen social ties and encourage collaborative efforts to address food security and environmental challenges.

Challenges and Considerations

While vertical farming offers numerous benefits, it also faces challenges that need to be addressed:

- High Initial Costs: The setup of vertical farms requires significant investment in infrastructure, technology, and energy systems. These high initial costs can be a barrier to entry for some entrepreneurs and organizations.

- Energy Consumption: Although vertical farms can reduce water and land use, they often require substantial energy for lighting, climate control, and water circulation. Transitioning to renewable energy sources can mitigate this issue but adds to the complexity and cost.

- Technical Expertise: Successful vertical farming operations require knowledge in agriculture, engineering, and environmental science. Recruiting and training skilled personnel is essential to manage and optimize these advanced systems effectively.

In conclusion, vertical farming is a transformative approach to urban agriculture that addresses many challenges associated with traditional farming. By maximizing space efficiency, reducing resource use, and providing fresh produce locally, vertical farming contributes to sustainable urban development. Despite the challenges, the environmental, economic, and social benefits make vertical farming a compelling solution for the future of urban food production.

Community Gardens

Community gardens are shared spaces where urban residents can collectively grow fruits, vegetables, herbs, and flowers. These gardens are typically managed by local organizations, schools, or neighborhood groups, and provide numerous benefits, including enhancing food security, fostering social cohesion, and promoting environmental stewardship. Community gardens play a vital role in urban agriculture, offering a practical and inclusive way to engage city dwellers in sustainable food production.

Design and Implementation

Establishing a community garden involves several steps, including securing land, organizing participants, and designing the garden layout. Key considerations include:

- Site Selection: Finding an appropriate location is crucial. Ideal sites have good sunlight, access to water, and healthy soil. Community gardens can be established on vacant lots, public parks, school grounds, or even rooftops.

- Planning and Design: The garden layout should accommodate the needs of all participants. Raised beds, individual plots, and communal growing areas are common features. Paths, composting areas, seating, and storage facilities should also be included.

- Organization and Governance: Effective management is essential for the success of a community garden. Establishing a governing body or committee to make decisions, allocate plots, and organize activities can help maintain order and ensure that the garden operates smoothly.

- Soil Preparation and Planting: Preparing the soil through testing and amendment is critical to ensure it is fertile and free from contaminants. Once the soil is ready, planting can begin. Participants typically decide what to grow based on local climate, soil conditions, and community preferences.

Environmental Benefits

Community gardens contribute to urban environmental sustainability in several ways:

- Biodiversity: These gardens enhance urban biodiversity by providing habitats for various plants, insects, birds, and other wildlife. The diversity of plants grown can attract pollinators and beneficial insects, contributing to a healthier urban ecosystem.

- Green Space: Community gardens transform unused or underutilized urban areas into green spaces, improving air quality, reducing the urban heat island effect, and contributing to the overall aesthetic of the neighborhood.

- Sustainable Practices: Many community gardens adopt organic and sustainable farming practices, avoiding chemical fertilizers and pesticides. This approach promotes soil health, reduces pollution, and encourages the recycling of organic waste through composting.

Social and Community Benefits

Community gardens foster a sense of community and social interaction, offering numerous social benefits:

- Community Engagement: These gardens provide a space for people of all ages and backgrounds to come together, work collaboratively, and share knowledge. This engagement strengthens social ties and builds a sense of belonging.

- Educational Opportunities: Community gardens serve as outdoor classrooms where participants can learn about gardening, nutrition, and sustainability. Schools and local organizations often use these gardens to teach children and adults about the importance of fresh food and environmental stewardship.

- Health and Well-being: Gardening is a physical activity that promotes exercise, reduces stress, and enhances mental health. Access to fresh, locally grown produce also improves nutrition and dietary habits, contributing to overall health and well-being.

Economic Benefits

Community gardens can provide economic advantages for urban residents and the wider community:

- Food Production: These gardens produce fresh, affordable produce, helping to reduce grocery bills for participants. This is particularly beneficial in low-income areas where access to fresh food may be limited.

- Job Creation and Skills Development: Community gardens can create job opportunities in gardening, education, and management. They also provide valuable skills training in horticulture, teamwork, and project management.

- Property Value and Urban Revitalization: The presence of well-maintained community gardens can increase property values and stimulate economic development. They can also revitalize neglected areas, making neighborhoods more attractive and livable.

Challenges and Considerations

While community gardens offer many benefits, they also face challenges that need to be addressed:

- Land Tenure: Securing long-term access to land can be difficult, especially in cities with high real estate values. Advocating for land trusts or securing leases with local governments can provide more stability.

- Funding and Resources: Community gardens often rely on donations, grants, and volunteer labor. Consistent funding is necessary for tools, seeds, infrastructure, and ongoing maintenance.

- Maintenance and Management: Effective management is crucial for the garden's success. This includes regular maintenance, conflict resolution among participants, and ensuring equitable access to garden resources.

In conclusion, community gardens are a vital component of urban agriculture, providing environmental, social, and economic benefits. By transforming urban spaces into productive green areas, community gardens enhance food security, foster community engagement, and promote sustainable living. Despite the challenges, the positive impact of community gardens on urban environments and residents makes them a valuable addition to cities worldwide.

Chapter 8: Policy and Economic Incentives

The successful implementation and scaling of nature-based solutions (NbS) in agriculture require robust policy frameworks and economic incentives. This chapter explores the role of government policies, international agreements, and financial mechanisms in promoting and supporting NbS. It examines the limitations of current agricultural policies and how targeted policy interventions can overcome these challenges. Additionally, the chapter discusses various subsidies and incentives that can encourage sustainable practices, the importance of supporting research and development, and the impact of international agreements and collaborations. Understanding these policy and economic drivers is crucial for creating an enabling environment that fosters the adoption of NbS and enhances the economic benefits of sustainable agriculture.

Current Agricultural Policies and Their Limitations

Agricultural policies play a crucial role in shaping farming practices, influencing land use, and determining the sustainability of agricultural systems. However, current agricultural policies often fall short in promoting sustainable practices and supporting the widespread adoption of NbS. This section examines the primary limitations of existing policies and highlights the need for comprehensive reforms to foster more sustainable agricultural systems.

Focus on Productivity Over Sustainability

Many agricultural policies prioritize maximizing crop yields and livestock production to ensure food security and economic growth. This focus often comes at the expense of environmental sustainability. Subsidies and incentives are frequently directed towards high-input, intensive farming practices that rely heavily on chemical fertilizers, pesticides, and monocropping. These practices can degrade soil health, reduce biodiversity, and contribute to water pollution and greenhouse gas emissions.

Inadequate Support for Small-Scale Farmers

Small-scale and subsistence farmers, who are crucial for local food security and rural livelihoods, often receive insufficient support from current agricultural policies. These farmers face numerous challenges, including limited access to credit, technology, and markets. Existing policies tend to favor large-scale industrial agriculture, leaving small-scale farmers without the necessary resources to adopt sustainable practices or invest in NbS.

Lack of Integration Across Sectors

Agricultural policies are frequently developed in isolation from other sectors, such as water management, forestry, and climate change. This lack of integration can lead to conflicting objectives and missed opportunities for synergistic solutions. For instance, policies that promote deforestation for agricultural expansion undermine efforts to conserve biodiversity and sequester carbon. An integrated approach that considers the interconnectedness of agriculture with other sectors is essential for achieving sustainable outcomes.

Insufficient Emphasis on Ecosystem Services

Ecosystem services, such as pollination, nutrient cycling, and soil formation, are vital for sustainable agriculture. However, current policies often fail to recognize and reward farmers for maintaining and enhancing these services. Payments for ecosystem services (PES) schemes, which provide financial incentives for practices that support ecosystem health, are not widely implemented. As a result, farmers lack motivation to adopt practices that protect and enhance natural resources.

Regulatory Barriers and Bureaucratic Hurdles

Complex and rigid regulatory frameworks can hinder the adoption of innovative and sustainable agricultural practices. Farmers may face bureaucratic challenges when trying to access subsidies, implement

NbS, or participate in conservation programs. Streamlining regulations and reducing administrative burdens can facilitate the transition to more sustainable farming systems.

Limited Investment in Research and Extension Services

The development and dissemination of sustainable agricultural technologies and practices require substantial investment in research and extension services. However, funding for agricultural research, particularly in the areas of sustainability and NbS, is often inadequate. Extension services that provide technical assistance and training to farmers are also underfunded, limiting their ability to support the adoption of sustainable practices.

Market Distortions

Market mechanisms often favor conventional agricultural products over those produced using sustainable methods. For example, externalities such as the environmental costs of intensive farming are not reflected in the prices of agricultural goods. This makes sustainably produced food less competitive in the market. Policies that internalize these externalities, such as carbon pricing or pollution taxes, can help level the playing field and promote sustainable agriculture.

In conclusion, current agricultural policies have significant limitations that hinder the widespread adoption of sustainable practices and NbS. Addressing these limitations requires a shift in policy focus towards sustainability, better support for small-scale farmers, integration across sectors, recognition of ecosystem services, streamlined regulations, increased investment in research and extension services, and market reforms. Comprehensive policy reforms are essential for fostering a sustainable and resilient agricultural sector that can meet the challenges of the 21st century.

Promoting NbS Through Policy Interventions

Promoting nature-based solutions (NbS) in agriculture requires targeted policy interventions that support sustainable practices and foster innovation. Effective policies can provide the necessary incentives, resources, and frameworks to encourage farmers to adopt NbS. This section explores how subsidies, research and development support, and international collaborations can drive the transition towards sustainable agriculture.

Subsidies and Incentives for Sustainable Practices

Subsidies and financial incentives are powerful tools that can encourage farmers to adopt sustainable practices and NbS. By shifting financial support towards environmentally friendly farming methods, governments can promote long-term sustainability and resilience in agriculture.

Redirection of Subsidies

Traditional agricultural subsidies often favor high-input, intensive farming practices that can harm the environment. Redirecting these subsidies towards sustainable practices can significantly impact farming methods. For example, subsidies could support organic farming, conservation tillage, cover cropping, and agroforestry. These practices enhance soil health, reduce reliance on chemical inputs, and increase biodiversity.

Payments for Ecosystem Services (PES)

Payments for Ecosystem Services (PES) programs compensate farmers for managing their land in ways that provide ecological benefits. These programs can reward farmers for practices that sequester carbon, enhance water quality, preserve habitats, and increase biodiversity. By financially recognizing the value of ecosystem services, PES programs create economic incentives for farmers to adopt NbS.

Tax Incentives

Tax incentives can also encourage sustainable agricultural practices. Governments can offer tax credits or deductions for expenses related to implementing NbS, such as purchasing cover crop seeds, installing efficient irrigation systems, or planting trees for agroforestry. These incentives reduce the financial burden on farmers and make sustainable practices more attractive.

Grants and Low-Interest Loans

Providing grants and low-interest loans for sustainable agricultural projects can help farmers invest in NbS. These financial instruments can fund the transition to organic farming, the establishment of riparian buffers, or the development of integrated crop-livestock systems. Access to affordable financing is crucial for farmers, especially small-scale producers, who may lack the capital to implement sustainable practices.

Market-Based Mechanisms

Market-based mechanisms, such as carbon markets and eco-certification schemes, can also promote NbS. Carbon markets allow farmers to sell carbon credits generated through practices like afforestation and reduced tillage, providing an additional revenue stream. Eco-certification schemes, such as organic or fair-trade labels, can increase market access and consumer demand for sustainably produced products, thereby incentivizing farmers to adopt NbS.

In summary, subsidies and incentives are essential for promoting sustainable agricultural practices and NbS. By redirecting financial support, offering PES programs, providing tax incentives, and facilitating access to grants and loans, governments can create an enabling environment that encourages farmers to adopt environmentally friendly practices.

Support for Research and Development

Research and development (R&D) play a critical role in advancing NbS and sustainable agriculture. Investment in R&D can lead to the development of new technologies, practices, and knowledge that support the widespread adoption of NbS.

Funding for Sustainable Agriculture Research

Governments should allocate substantial funding for research focused on sustainable agriculture and NbS. This includes studies on soil health, water management, crop diversity, pest control, and climate resilience. Research institutions, universities, and agricultural extension services can collaborate to develop innovative solutions that address the specific needs of different regions and farming systems.

Development of New Technologies

Technological innovation is essential for enhancing the efficiency and effectiveness of NbS. Investment in the development of new technologies, such as precision agriculture tools, advanced irrigation systems, and biocontrol agents, can help farmers implement sustainable practices. These technologies can improve resource use efficiency, reduce environmental impact, and increase productivity.

Knowledge Transfer and Extension Services

Effective knowledge transfer is crucial for the adoption of NbS. Agricultural extension services play a vital role in disseminating research findings and providing technical assistance to farmers. Strengthening these services and ensuring they have the resources and expertise to promote sustainable practices can accelerate the adoption of NbS. Training programs, field demonstrations, and farmer-to-farmer learning initiatives can also enhance knowledge transfer.

Public-Private Partnerships

Collaboration between the public and private sectors can drive innovation in sustainable agriculture. Public-private partnerships can leverage the strengths of both sectors, combining public research capabilities with private sector investment and market reach. These partnerships can develop and commercialize new technologies, create sustainable value chains, and support smallholder farmers.

Data and Monitoring Systems

Robust data collection and monitoring systems are essential for evaluating the effectiveness of NbS and guiding policy decisions. Investment in remote sensing, geographic information systems (GIS), and other monitoring technologies can provide valuable insights into land use, crop performance, and environmental impact. These data can inform adaptive management strategies and ensure that NbS deliver the desired outcomes.

In conclusion, supporting research and development is vital for advancing NbS in agriculture. By funding sustainable agriculture research, developing new technologies, enhancing knowledge transfer, fostering public-private partnerships, and investing in data and monitoring systems, governments can promote innovation and the adoption of sustainable practices.

International Agreements and Collaborations

International agreements and collaborations are crucial for addressing global agricultural challenges and promoting NbS. These agreements facilitate knowledge exchange, coordinate actions, and mobilize resources to support sustainable agriculture worldwide.

Global Frameworks and Agreements

International frameworks, such as the Paris Agreement on climate change and the Convention on Biological Diversity, provide a foundation for collaborative efforts to promote NbS. These agreements set global targets for greenhouse gas emissions reduction, biodiversity conservation, and sustainable land use. By aligning national policies with these frameworks, countries can contribute to global sustainability goals and benefit from international support and cooperation.

Regional Collaborations

Regional collaborations, such as the European Union's Common Agricultural Policy (CAP) and the African Union's Comprehensive Africa Agriculture Development Programme (CAADP), can enhance the implementation of NbS. These collaborations provide platforms for countries to share best practices, harmonize policies, and coordinate efforts to address regional agricultural challenges. Joint initiatives can leverage regional strengths and resources to promote sustainable agriculture.

Technical Assistance and Capacity Building

International organizations, such as the Food and Agriculture Organization (FAO) and the International Fund for Agricultural Development (IFAD), play a key role in providing technical assistance and capacity building for NbS. These organizations offer training programs, technical expertise, and funding to help countries develop and implement sustainable agricultural practices. Capacity building initiatives can strengthen national institutions, enhance policy frameworks, and empower farmers to adopt NbS.

Research and Knowledge Exchange

International research collaborations and knowledge exchange platforms, such as the Global Alliance for Climate-Smart Agriculture (GACSA) and the Consultative Group on International Agricultural Research (CGIAR), facilitate the sharing of scientific knowledge and innovative practices. These collaborations support joint research projects, data sharing, and the dissemination of findings to policymakers and practitioners. By fostering a global community of practice, these initiatives accelerate the adoption of NbS.

Financial Mechanisms

International financial mechanisms, such as the Green Climate Fund (GCF) and the Global Environment Facility (GEF), provide funding for projects that promote NbS and sustainable agriculture. These funds support initiatives that enhance climate resilience, conserve biodiversity, and improve land and water management. Access to international funding can help countries overcome financial barriers and scale up NbS implementation.

In summary, international agreements and collaborations are essential for promoting NbS in agriculture. By participating in global frameworks, engaging in regional collaborations, leveraging technical assistance, facilitating research exchange, and accessing international funding, countries can enhance their capacity to implement sustainable agricultural practices and contribute to global sustainability goals.

Economic Benefits of Nature-Based Solutions

NbS offer numerous economic benefits that extend beyond environmental sustainability. By integrating natural processes into agricultural practices, NbS can enhance productivity, reduce costs, create new market opportunities, and support resilient rural economies. This section explores the economic advantages of adopting NbS in agriculture.

Increased Agricultural Productivity

Implementing NbS can lead to higher agricultural productivity through improved soil health, enhanced water management, and increased biodiversity. Practices such as agroforestry, cover cropping, and conservation tillage enhance soil structure and fertility, leading to better crop yields. For instance, agroforestry systems integrate trees with crops and livestock, providing multiple outputs such as timber, fruits, and fodder, in addition to the primary crop. This diversification not only increases overall productivity but also spreads risk, ensuring more stable income for farmers.

Cost Savings

NbS can reduce the need for expensive agricultural inputs such as synthetic fertilizers, pesticides, and irrigation water. Sustainable practices like crop rotation, integrated pest management (IPM), and the use of organic amendments improve soil fertility and pest control naturally, reducing the dependence on chemical inputs. For example, IPM leverages natural predators to manage pest populations, lowering the cost of chemical pesticides. Similarly, practices that enhance soil organic matter, such as composting and cover cropping, reduce the need for synthetic fertilizers by improving nutrient availability and soil health.

Enhanced Resilience to Climate Change

NbS enhance the resilience of agricultural systems to climate change, thereby reducing economic losses associated with extreme weather events. Practices like agroforestry and riparian buffers help mitigate the impacts of droughts, floods, and storms by stabilizing soil, improving water infiltration, and reducing surface runoff. This resilience translates into more consistent yields and reduced crop failures, ensuring a more reliable income for farmers. Additionally, by sequestering carbon and reducing greenhouse gas emissions, NbS contribute to climate mitigation efforts, potentially providing farmers with additional revenue streams through carbon credits.

Creation of New Market Opportunities

The adoption of NbS can open up new market opportunities for farmers, particularly in niche markets for organic and sustainably produced products. Consumers are increasingly willing to pay a premium for products that are certified organic, fair trade, or sustainably sourced. NbS practices such as organic farming, agroforestry, and permaculture align well with these market trends, allowing farmers to access higher-value markets and increase their profit margins. Additionally, NbS can create opportunities for eco-tourism and educational services, where farms serve as destinations for visitors interested in learning about sustainable agriculture.

Job Creation and Rural Development

Implementing NbS can stimulate rural economies by creating jobs and supporting local businesses. The adoption of sustainable practices often requires skilled labor for activities such as planting and maintaining trees, managing diverse cropping systems, and operating new technologies. This demand for labor can create employment opportunities in rural areas, contributing to poverty reduction and economic development. Moreover, the development of local value chains for sustainably produced goods can support small and medium-sized enterprises, fostering economic diversification and resilience in rural communities.

Reduction of External Costs

NbS reduce external costs associated with conventional agriculture, such as environmental degradation, health impacts from pesticide use, and loss of ecosystem services. By enhancing ecosystem health, NbS provide valuable services like water purification, pollination, and climate regulation, which would otherwise require costly technical solutions. These external cost savings benefit society as a whole and justify public investments in NbS through subsidies, grants, and supportive policies.

Long-Term Economic Sustainability

NbS promote long-term economic sustainability by ensuring the continued availability of natural resources essential for agriculture. Practices that protect soil health, conserve water, and maintain biodiversity ensure that agricultural systems remain productive and resilient over the long term. This sustainability reduces the risk of resource depletion and environmental collapse, safeguarding future food security and economic stability.

In conclusion, nature-based solutions offer a wide range of economic benefits that enhance productivity, reduce costs, create new market opportunities, and support resilient rural economies. By promoting sustainable agricultural practices, NbS contribute to both environmental health and economic prosperity, making them a vital component of a sustainable and resilient agricultural future.

Chapter 9: Farmer and Community Engagement

Effective engagement of farmers and local communities is crucial for the successful implementation and sustainability of nature-based solutions (NbS) in agriculture. This chapter examines the importance of involving key stakeholders, including farmers, local communities, and non-governmental organizations (NGOs), in the decision-making and implementation processes. It also explores strategies for fostering meaningful engagement through participatory approaches, knowledge sharing, education, and building community resilience. By prioritizing stakeholder involvement, we can ensure that NbS are tailored to local needs, widely accepted, and sustainably maintained.

Importance of Stakeholder Involvement

Engaging stakeholders, particularly farmers, local communities, and non-governmental organizations (NGOs), is essential for the successful implementation and sustainability of nature-based solutions (NbS) in agriculture. Stakeholder involvement ensures that the practices adopted are relevant, accepted, and maintained over the long term. This section delves into the critical roles that farmers, local communities, and NGOs play in promoting and sustaining NbS.

Farmers, Local Communities, and NGOs

Farmers, local communities, and NGOs play pivotal roles in the successful implementation and sustainability of NbS in agriculture.

Farmers

Farmers are the primary custodians of agricultural lands and play a pivotal role in the adoption and implementation of NbS. Their involvement is crucial for several reasons:

- Knowledge and Experience: Farmers possess extensive knowledge and experience regarding their land, crops, and local environmental conditions. This expertise is invaluable for identifying suitable NbS practices that align with local agroecological conditions. Engaging farmers in the planning and decision-making process ensures that the solutions proposed are practical and tailored to their specific needs.

- Adoption and Maintenance: The successful adoption of NbS relies heavily on the willingness and ability of farmers to implement and maintain these practices. When farmers are actively involved in the development and implementation of NbS, they are more likely to take ownership and responsibility for the long-term success of these initiatives. This commitment is essential for the sustainability of NbS.

- Innovation and Adaptation: Farmers are often innovative and capable of adapting practices to suit their unique contexts. Involving them in NbS projects encourages the exchange of ideas and the adaptation of techniques, leading to more effective and resilient solutions. Farmers can also provide feedback on the effectiveness of NbS, facilitating continuous improvement and refinement of practices.

Local Communities

The engagement of local communities is equally important for the success of NbS. Community involvement provides several benefits:

- Social Acceptance and Support: NbS that involve and benefit local communities are more likely to gain social acceptance and support. Community members who see tangible benefits, such as improved food security, enhanced livelihoods, and better environmental conditions, are more likely to advocate for and support NbS initiatives.

- Collective Action: Many NbS, such as watershed management, reforestation, and community gardens, require collective action and collaboration. Engaging the community fosters a sense of shared responsibility and cooperation, which is essential for the success of these projects. Collective action also enhances the scale and impact of NbS, leading to broader environmental and social benefits.

- Resilience and Empowerment: Involving local communities in NbS projects empowers them to take control of their environmental and economic futures. This empowerment builds community resilience, as members are better equipped to manage natural resources sustainably and respond to environmental challenges. Additionally, community-driven NbS can provide opportunities for education, skill development, and income generation, further strengthening community resilience.

Non-Governmental Organizations

NGOs play a crucial intermediary role in promoting and supporting NbS. Their involvement offers several advantages:

- Capacity Building: NGOs often have the resources and expertise to provide training, technical assistance, and capacity-building support to farmers and communities. This support is vital for the successful implementation and scaling of NbS, as it equips stakeholders with the knowledge and skills needed to adopt and maintain sustainable practices.

- Advocacy and Policy Influence: NGOs can advocate for policies and regulations that support NbS and sustainable agriculture. By working with governments, NGOs can help create an enabling policy environment that incentivizes the adoption of NbS and addresses barriers to implementation.

- Resource Mobilization: NGOs can mobilize financial and technical resources to support NbS projects. Through fundraising, grant applications, and partnerships, NGOs can secure the necessary funding and expertise to implement NbS at scale. Their ability to connect local projects with international funding and expertise enhances the impact and sustainability of NbS initiatives.

In conclusion, the involvement of farmers, local communities, and NGOs is critical for the successful implementation and sustainability of NbS in agriculture. By leveraging the knowledge, experience, and resources of these stakeholders, NbS projects can be more effectively designed, adopted, and maintained, leading to long-term environmental and social benefits.

Strategies for Effective Engagement

Effective engagement of stakeholders is essential for the successful implementation and sustainability of NbS in agriculture. Strategies that prioritize participation, knowledge sharing, education, and community resilience can foster meaningful involvement of farmers, local communities, and NGOs. This section explores various approaches to ensure stakeholders are actively engaged and empowered to contribute to the adoption and success of NbS.

Participatory Approaches

Participatory approaches in agriculture involve engaging farmers, local communities, and other stakeholders in the planning, implementation, and evaluation of projects. These approaches are essential for promoting NbS as they ensure that interventions are tailored to local needs, culturally appropriate, and widely accepted. By fostering active involvement and collaboration, participatory approaches can enhance the effectiveness and sustainability of NbS in agricultural settings.

Community-Led Planning

Community-led planning is a participatory approach where local communities take the lead in identifying problems, setting priorities, and designing interventions. This process begins with community meetings and workshops where stakeholders can voice their concerns, share their knowledge, and contribute to decision-making. Facilitators, often from NGOs or government agencies, guide these discussions to ensure that all voices are heard and considered.

- Benefits: This approach ensures that the interventions are relevant to the community's specific context and needs. It also fosters a sense of ownership and responsibility among community members, increasing their commitment to the project's success.

- Challenges: Effective community-led planning requires skilled facilitators who can manage diverse perspectives and potential conflicts. It also demands time and resources to build trust and consensus within the community.

Participatory Rural Appraisal (PRA)

Participatory Rural Appraisal (PRA) is a set of methods and tools used to involve communities in the analysis of their situation and the planning of development activities. PRA techniques include mapping, transect walks, seasonal calendars, and focus group discussions. These tools help communities gather and analyze information about their environment, resources, and livelihoods.

- Benefits: PRA methods are interactive and inclusive, allowing community members to contribute their knowledge and insights. They help identify local solutions and build community capacity to manage projects.

- Challenges: PRA requires trained facilitators to guide the process and ensure that it remains participatory. There is also a risk of bias if certain groups dominate the discussions or if external facilitators impose their own agendas.

Farmer Field Schools

Farmer Field Schools (FFS) are participatory extension approaches that involve farmers in learning and experimenting with new practices in a collaborative setting. FFS groups meet regularly to observe, discuss, and experiment with NbS techniques such as integrated pest management, soil conservation, and water management. The learning process is experiential and based on field observations and practical activities.

- Benefits: FFS empower farmers by enhancing their knowledge and skills through hands-on learning. They promote peer-to-peer learning and foster a community of practice that can sustain NbS adoption.

- Challenges: Establishing and maintaining FFS requires resources and skilled facilitators. Ensuring continuity and scaling up successful practices beyond the initial groups can also be challenging.

Participatory Monitoring and Evaluation (PM&E)

Participatory Monitoring and Evaluation (PM&E) involves stakeholders in the tracking and assessment of project progress and outcomes. Community members help define indicators, collect data, and analyze results. This approach ensures that monitoring and evaluation are grounded in local realities and provide feedback for adaptive management.

- Benefits: PM&E promotes transparency and accountability, ensuring that projects remain responsive to community needs. It also builds local capacity for self-assessment and continuous improvement.

- Challenges: PM&E can be resource-intensive and time-consuming. It requires training for community members to effectively participate in data collection and analysis.

Inclusive Decision-Making

Inclusive decision-making ensures that all community members, including marginalized groups such as women, youth, and indigenous peoples, have a voice in project planning and implementation. Strategies to promote inclusivity include holding meetings at convenient times, providing translation services, and creating safe spaces for participation.

- Benefits: Inclusive decision-making leads to more equitable and just outcomes. It ensures that NbS address the needs and priorities of all community members, enhancing social cohesion and support for the project.

- Challenges: Achieving genuine inclusivity can be difficult, especially in communities with entrenched social hierarchies and power imbalances. It requires continuous effort and sensitivity to local dynamics.

Collaborative Partnerships

Forming collaborative partnerships with local institutions, NGOs, research organizations, and government agencies can enhance the implementation of participatory approaches. These partnerships bring additional resources, expertise, and legitimacy to NbS projects.

- Benefits: Partnerships can provide technical support, funding, and policy advocacy, strengthening the overall impact and sustainability of NbS initiatives. They also facilitate knowledge exchange and innovation.

- Challenges: Managing partnerships requires effective communication and coordination. Conflicting interests and priorities among partners can pose challenges to collaboration.

In conclusion, participatory approaches are crucial for the successful implementation of nature-based solutions in agriculture. By involving farmers, local communities, and other stakeholders in the planning, execution, and evaluation of projects, participatory approaches ensure that NbS are relevant, accepted, and sustainable. While these approaches come with challenges, their benefits in terms of local ownership, capacity building, and community resilience make them indispensable for sustainable agricultural development.

Knowledge Sharing and Education

Knowledge sharing and education are vital components for the successful adoption and sustainability of NbS in agriculture. These processes ensure that farmers and communities have access to the information, skills, and resources needed to implement and maintain sustainable practices. By fostering a culture of continuous learning and collaboration, knowledge sharing and education can drive innovation and improve the resilience of agricultural systems.

Extension Services and Training Programs

Extension services and training programs play a crucial role in disseminating knowledge about NbS. These services are often provided by government agencies, NGOs, universities, and research institutions. They offer farmers practical advice, technical support, and hands-on training to adopt and adapt NbS.

- Workshops and Field Days: Organizing workshops and field days where farmers can see NbS in action and interact with experts is an effective way to transfer knowledge. These events provide opportunities for farmers to ask questions, share experiences, and learn from each other.

- Demonstration Projects: Establishing demonstration projects that showcase successful NbS practices can serve as learning hubs for farmers. These projects illustrate the practical benefits and feasibility of NbS, encouraging wider adoption.

- On-Farm Trials: Involving farmers in on-farm trials allows them to experiment with NbS practices on their own land. This participatory approach helps farmers understand the benefits and challenges of NbS firsthand and promotes local innovation.

Farmer-to-Farmer Learning

Farmer-to-farmer learning is a powerful method for spreading knowledge about NbS. Peer learning leverages the trust and credibility that farmers have with each other, facilitating the exchange of practical and context-specific knowledge.

- Farmer Networks and Cooperatives: Establishing farmer networks and cooperatives can provide platforms for knowledge sharing. These networks enable farmers to share experiences, resources, and best practices related to NbS.

- Study Tours and Exchange Visits: Organizing study tours and exchange visits where farmers can visit other farms that have successfully implemented NbS can inspire and motivate them to adopt similar practices. Seeing the tangible benefits of NbS in different contexts helps farmers envision how they can apply these solutions on their own farms.

- Mentorship Programs: Pairing experienced farmers who have successfully adopted NbS with those who are new to these practices can provide ongoing support and guidance. Mentorship programs foster a collaborative learning environment and build farmer capacity over time.

Use of Digital Tools and Media

Digital tools and media can significantly enhance knowledge sharing and education efforts. These tools provide farmers with access to information and training resources, regardless of their location.

- Online Platforms and Mobile Applications: Creating online platforms and mobile applications that provide information on NbS, including guidelines, best practices, and troubleshooting tips, can support farmers in implementing sustainable practices. These platforms can also facilitate virtual training sessions and webinars.

- Social Media and Messaging Groups: Leveraging social media and messaging platforms to create farmer groups and communities can enhance communication and knowledge sharing. Farmers can use these groups to ask questions, share experiences, and receive real-time support from peers and experts.

- Radio and Television Programs: Broadcasting educational programs about NbS on radio and television can reach a broad audience, including those with limited access to digital tools. These programs can feature success stories, expert interviews, and practical advice on implementing NbS.

Educational Institutions and Curricula

Integrating NbS into the curricula of agricultural education institutions ensures that future farmers and agricultural professionals are equipped with the knowledge and skills needed to implement sustainable practices.

- Universities and Technical Schools: Developing courses and degree programs focused on sustainable agriculture and NbS can prepare students for careers in this field. Hands-on training, internships, and research projects can provide practical experience.

- School Gardens and Educational Programs: Establishing school gardens and educational programs that teach children about NbS and sustainable agriculture can foster an early appreciation for these practices. Engaging students in gardening activities helps them understand the importance of protecting natural resources and adopting sustainable practices.

Participatory Research and Citizen Science

Engaging farmers and community members in participatory research and citizen science initiatives can enhance knowledge sharing and education. These approaches involve stakeholders in the research process, from identifying research questions to collecting and analyzing data.

- Participatory Research Projects: Collaborating with farmers to conduct research on NbS practices ensures that the research is relevant and applicable to their needs. This collaboration can also build local capacity for research and problem-solving.

- Citizen Science Initiatives: Encouraging community members to participate in data collection and monitoring efforts can increase their understanding of NbS and their impacts. Citizen science projects can generate valuable data for researchers and policymakers while empowering participants with knowledge and skills.

In conclusion, knowledge sharing and education are essential for promoting the adoption and sustainability of nature-based solutions in agriculture. Through extension services, farmer-to-farmer learning, digital tools, educational institutions, and participatory research, stakeholders can access the information and support needed to implement and maintain sustainable practices. By fostering a culture of continuous learning and collaboration, these efforts can drive innovation and enhance the resilience of agricultural systems.

Building Community Resilience

Building community resilience is crucial for the sustainable implementation and long-term success of NbS in agriculture. Resilient communities are better equipped to adapt to environmental, social, and economic changes, ensuring that NbS can thrive under varying conditions. This section explores key strategies for fostering community resilience, focusing on empowering local communities, enhancing social cohesion, and promoting sustainable livelihoods.

Empowering Local Communities

Empowering local communities involves providing them with the knowledge, skills, and resources needed to implement and sustain NbS. This empowerment fosters a sense of ownership and responsibility, which is critical for the resilience and success of NbS projects.

- Capacity Building: Offering training and capacity-building programs helps community members develop the technical skills required for NbS. These programs can include workshops on sustainable farming practices, water management, and soil conservation, as well as training in financial management and project planning.

- Access to Resources: Ensuring that communities have access to necessary resources, such as seeds, tools, and financial support, is vital for the successful implementation of NbS. Microcredit schemes, grants, and subsidies can provide the financial backing needed to invest in sustainable practices.

- Leadership Development: Encouraging and nurturing local leadership is essential for building resilience. Leadership development programs can identify and train community leaders who can champion NbS initiatives, mobilize resources, and inspire others to participate.

Enhancing Social Cohesion

Strong social cohesion within communities enhances their ability to collaborate and support one another in implementing NbS. Social networks and trust are critical components of resilience, enabling communities to respond collectively to challenges and opportunities.

- Community Engagement: Actively engaging community members in the planning, decision-making, and implementation processes fosters a sense of belonging and shared purpose. Community meetings, participatory workshops, and collaborative projects can facilitate this engagement.

- Building Networks: Creating and strengthening networks among community members, farmers, local organizations, and external stakeholders enhances knowledge exchange and mutual support. These networks can provide platforms for sharing best practices, troubleshooting issues, and coordinating efforts.

- Cultural Integration: Integrating NbS into local cultural practices and values ensures their acceptance and sustainability. Recognizing and respecting traditional knowledge and practices can enhance the relevance and effectiveness of NbS.

Promoting Sustainable Livelihoods

Ensuring that NbS contribute to sustainable livelihoods is essential for building community resilience. Economic stability and diversification reduce vulnerability to shocks and stresses, enabling communities to invest in and maintain NbS.

- Income Diversification: Promoting diversified income sources through NbS can enhance economic resilience. For example, agroforestry systems that produce fruits, nuts, timber, and non-timber forest products can provide multiple revenue streams. Similarly, integrating livestock with crop production can offer additional income and food security.

- Value Addition: Encouraging value addition through processing and marketing of NbS products can increase profitability and create local employment opportunities. Training in value-added processing, branding, and market access can help communities capture higher returns from their products.

- Sustainable Market Linkages: Establishing sustainable market linkages for NbS products ensures stable demand and fair prices.

Supporting farmers and communities in accessing local, regional, and international markets can enhance the economic viability of NbS. Fair trade certifications and eco-labeling can also help communities secure better market opportunities.

Adaptive Management

Adaptive management practices enable communities to respond flexibly to changing conditions and uncertainties. By continuously monitoring and adjusting NbS strategies, communities can maintain their effectiveness and relevance.

- Monitoring and Evaluation: Implementing robust monitoring and evaluation systems allows communities to track the progress and impact of NbS. Regular assessments help identify challenges and opportunities for improvement, ensuring that NbS remain effective and beneficial.

- Feedback Mechanisms: Establishing feedback mechanisms enables community members to voice their concerns, share experiences, and suggest improvements. This participatory approach ensures that NbS are continuously refined and adapted to local needs and conditions.

- Scenario Planning: Engaging in scenario planning helps communities anticipate and prepare for potential future challenges. By considering various scenarios, such as climate change impacts or market fluctuations, communities can develop strategies to mitigate risks and enhance resilience.

In conclusion, building community resilience is essential for the successful and sustainable implementation of nature-based solutions in agriculture. Empowering local communities, enhancing social cohesion, promoting sustainable livelihoods, and adopting adaptive management practices are key strategies for fostering resilience. By focusing on these areas, communities can create robust and adaptable systems that support the long-term success of NbS and contribute to overall sustainability and well-being.

Chapter 10: Future Directions and Innovations

As the need for sustainable agricultural practices becomes increasingly urgent, the future of nature-based solutions (NbS) in agriculture looks promising. This chapter explores emerging trends and innovations that are driving the adoption and effectiveness of NbS. It delves into technological advancements, new research developments, and the potential for scaling up and mainstreaming these solutions. Additionally, the chapter discusses the role of international collaboration in overcoming barriers to adoption and envisions a future where sustainable agriculture is the norm. By examining these future directions, we can better understand how to foster resilient and productive agricultural systems that benefit both people and the planet.

Emerging Trends in Nature-Based Solutions

NbS are gaining traction as essential components of sustainable agricultural systems. These solutions harness natural processes to address environmental challenges, enhance resilience, and improve productivity. As we look to the future, several emerging trends are set to shape the evolution and adoption of NbS. This section explores the role of technological innovations and new research developments in advancing NbS and ensuring their widespread implementation.

Technological Innovations

Technological innovations are revolutionizing the way nature-based solutions are implemented and managed in agriculture. These advancements enhance the efficiency, effectiveness, and scalability of NbS, making them more accessible to farmers and communities worldwide.

Precision Agriculture

Precision agriculture involves the use of advanced technologies to optimize farming practices based on detailed, site-specific data. Tools such as GPS, remote sensing, and geographic information systems (GIS) enable farmers to monitor and manage their fields with high precision. For NbS, precision agriculture can optimize the placement and management of cover crops, agroforestry systems, and buffer zones, ensuring maximum environmental and economic benefits.

- Benefits: Precision agriculture reduces input use (such as water, fertilizers, and pesticides), minimizes environmental impact, and increases crop yields. It allows for precise implementation of NbS, enhancing their effectiveness and sustainability.

- Challenges: The initial cost of precision agriculture technologies can be high, and there is a need for training and technical support to ensure farmers can effectively use these tools.

Drones and Remote Sensing

Drones equipped with multispectral and thermal cameras provide real-time data on crop health, soil conditions, and water stress. This information is crucial for implementing and monitoring NbS, such as precision irrigation and pest management.

- Benefits: Drones offer detailed and accurate data that can guide decision-making and improve the management of NbS. They enable early detection of issues, allowing for timely interventions that enhance resilience and productivity.

- Challenges: Ensuring data accuracy and integrating drone technology into existing farming practices can be complex. Additionally, regulatory frameworks for drone use in agriculture are still evolving.

Artificial Intelligence (AI) and Machine Learning

AI and machine learning algorithms analyze vast amounts of data to identify patterns and make predictions. These technologies can optimize NbS by predicting pest outbreaks, recommending optimal planting times, and assessing the impact of different management practices.

- Benefits: AI and machine learning enhance decision-making, reduce risks, and increase the efficiency of NbS. They can process complex datasets to provide actionable insights, supporting adaptive management.

- Challenges: Developing and training AI models requires substantial data and computational resources. Ensuring the accuracy and reliability of AI predictions is also crucial.

Biotechnologies

Biotechnologies, such as genetic engineering and synthetic biology, offer new opportunities for enhancing NbS. For example, developing crop varieties that are more resilient to environmental stresses or capable of fixing nitrogen can reduce the need for chemical inputs and support sustainable farming systems.

- Benefits: Biotechnologies can improve crop performance, enhance resilience to climate change, and reduce the environmental footprint of agriculture. They offer innovative solutions for integrating NbS into modern farming systems.

- Challenges: The use of biotechnologies raises ethical and regulatory concerns, and there is a need for careful assessment of their potential risks and benefits.

New Research and Developments

Ongoing research and new developments are continually expanding our understanding of NbS and how to implement them effectively.

These advancements are critical for refining NbS practices, overcoming challenges, and scaling their adoption.

Soil Health and Microbiomes

Research on soil health and microbiomes is uncovering the complex interactions between plants, soil organisms, and the environment. Understanding these interactions is essential for optimizing NbS that enhance soil fertility, such as cover cropping, composting, and reduced tillage.

- Benefits: Improved knowledge of soil microbiomes can lead to better soil management practices, increased crop productivity, and enhanced ecosystem services. It supports the development of tailored NbS that meet specific soil health needs.

- Challenges: Translating scientific findings into practical recommendations for farmers can be challenging. There is a need for effective communication and extension services to bridge this gap.

Climate Resilience

Research is focusing on developing NbS that enhance the resilience of agricultural systems to climate change. This includes breeding climate-resilient crop varieties, designing agroforestry systems that buffer against extreme weather, and developing water management strategies that mitigate drought and flood risks.

- Benefits: Climate-resilient NbS can help farmers adapt to changing conditions, ensuring food security and sustainable livelihoods. These solutions contribute to both mitigation and adaptation efforts.

- Challenges: Developing and implementing climate-resilient NbS requires interdisciplinary collaboration and long-term research investment. Farmers need support to adopt these practices and integrate them into their existing systems.

Biodiversity and Ecosystem Services

Research on biodiversity and ecosystem services is highlighting the critical role of diverse plant and animal communities in supporting sustainable agriculture. Studies are exploring how NbS such as pollinator habitats, hedgerows, and buffer strips enhance biodiversity and provide ecosystem services that benefit farming.

- Benefits: Biodiversity-focused NbS can improve pollination, pest control, and soil health, leading to more productive and resilient agricultural systems. They also contribute to broader conservation goals.

- Challenges: Measuring and valuing ecosystem services can be complex, and there is a need for standardized metrics and assessment tools. Encouraging farmers to prioritize biodiversity alongside production goals requires effective incentives and policy support.

Social and Economic Impacts

Understanding the social and economic impacts of NbS is crucial for their widespread adoption. Research is examining how NbS affect rural livelihoods, equity, and community resilience. This includes assessing the economic viability of NbS, identifying barriers to adoption, and exploring strategies for scaling up successful practices.

- Benefits: Insights into the social and economic dimensions of NbS can inform policies and programs that support equitable and sustainable agricultural development. They ensure that NbS contribute to inclusive growth and social well-being.

- Challenges: Addressing the diverse needs and contexts of different farming communities requires flexible and context-specific approaches. Ensuring that NbS are accessible and beneficial to all farmers, including smallholders and marginalized groups, is essential.

In conclusion, emerging trends in technological innovations and new research developments are driving the advancement of nature-based solutions in agriculture. These trends enhance the efficiency, effectiveness, and scalability of NbS, ensuring that they can contribute to sustainable and resilient agricultural systems. By staying at the forefront of these innovations and incorporating new knowledge, stakeholders can support the widespread adoption and success of NbS.

Scaling Up and Mainstreaming NbS

Scaling up and mainstreaming NbS in agriculture is essential for achieving widespread environmental, social, and economic benefits. Despite the clear advantages of NbS, several barriers hinder their broader adoption. Addressing these barriers through strategic interventions and leveraging international collaboration can significantly enhance the uptake and impact of NbS. This section explores the challenges to adoption and the critical role of international cooperation in promoting NbS.

Overcoming Barriers to Adoption

Adopting nature-based solutions in agriculture faces several challenges that need to be addressed to ensure widespread implementation. These barriers range from financial constraints to knowledge gaps and policy limitations. Identifying and overcoming these barriers is crucial for mainstreaming NbS.

1. Financial Constraints:

- High Initial Costs: The upfront costs of implementing NbS, such as establishing agroforestry systems or installing efficient irrigation, can be prohibitive for many farmers, especially smallholders.

- Limited Access to Credit: Farmers often lack access to affordable credit or financial products tailored to sustainable agriculture investments. Financial institutions may perceive NbS as risky or unprofitable.

- Solutions: Governments and financial institutions can develop and offer tailored financial products, such as low-interest loans, grants, and subsidies, specifically for NbS. Additionally, payment for ecosystem services (PES) programs can provide farmers with financial incentives to adopt sustainable practices.

2. Knowledge and Skills Gaps:

- Lack of Awareness: Many farmers are unaware of the benefits and methods of NbS. Misinformation or lack of information can lead to resistance to change.

- Technical Skills: Implementing NbS requires specific technical knowledge and skills that many farmers do not possess.

- Solutions: Extensive training and extension services are essential. Governments, NGOs, and research institutions can organize workshops, field days, and farmer-to-farmer learning programs to disseminate knowledge and build skills. Demonstration projects and success stories can also help illustrate the benefits of NbS.

3. Policy and Regulatory Barriers:

- Inadequate Policies: Existing agricultural policies may not support NbS or may actively promote unsustainable practices through subsidies and incentives for high-input farming.

- Regulatory Hurdles: Complex and restrictive regulations can hinder the adoption of NbS, especially when they involve land use changes or new farming practices.

- Solutions: Policy reforms are necessary to create an enabling environment for NbS. This includes revising agricultural subsidies to favor sustainable practices, simplifying regulatory processes, and incorporating NbS into national agricultural and environmental policies.

4. Market Access and Incentives:

- Market Limitations: Farmers adopting NbS may face challenges in accessing markets for their products, especially if these products are new or niche.

- Price Competitiveness: Sustainable products may have higher production costs, making them less competitive compared to conventional products.

- Solutions: Developing value chains for sustainably produced goods, creating market linkages, and promoting certification schemes (e.g., organic, fair trade) can enhance market access. Government procurement policies can also prioritize sustainably produced goods, creating demand and supporting price competitiveness.

5. Social and Cultural Barriers:

- Resistance to Change: Traditional farming practices are deeply rooted in many communities. Social norms and cultural values can resist changes, even if those changes are beneficial.

- Community Dynamics: Power dynamics within communities can affect the adoption of NbS, with influential individuals either supporting or resisting new practices.

- Solutions: Engaging communities through participatory approaches ensures that NbS are culturally appropriate and socially accepted.

Building trust, demonstrating respect for traditional knowledge, and involving community leaders can facilitate acceptance and adoption.

In summary, addressing financial constraints, knowledge gaps, policy barriers, market access issues, and social resistance is essential for scaling up and mainstreaming NbS in agriculture. Targeted interventions and support from governments, financial institutions, and NGOs can help overcome these challenges, paving the way for widespread adoption of sustainable practices.

The Role of International Collaboration

International collaboration plays a vital role in promoting the adoption and scaling of nature-based solutions in agriculture. By leveraging global networks, sharing knowledge, and coordinating efforts, countries can enhance the effectiveness and reach of NbS. This section explores how international cooperation can address barriers to adoption and support sustainable agricultural practices.

1. Knowledge Exchange and Capacity Building:

- Global Knowledge Networks: International collaboration facilitates the exchange of knowledge and best practices across countries. Platforms such as the Global Alliance for Climate-Smart Agriculture (GACSA) and the Food and Agriculture Organization (FAO) provide valuable resources and forums for sharing experiences with NbS.

- Training Programs: Collaborative training programs and workshops can build the capacity of farmers, extension workers, and policymakers. These programs can be organized at regional and international levels to disseminate technical knowledge and skills related to NbS.

- Solutions: Establishing international partnerships between research institutions, universities, and agricultural organizations can enhance knowledge sharing. Joint research projects and field trials can generate valuable data and insights, which can be shared globally through publications, conferences, and online platforms.

2. Policy Harmonization and Advocacy:

- Coordinated Policy Frameworks: International cooperation can help harmonize policies and regulations related to NbS. Countries can learn from each other's policy successes and challenges, adopting and adapting effective measures to their own contexts.

- Advocacy for Supportive Policies: International organizations can advocate for policies that support NbS at the global level. This includes promoting sustainable agriculture in international agreements, such as the Paris Agreement on climate change and the Convention on Biological Diversity.

- Solutions: Developing regional policy frameworks that align with global sustainability goals can enhance the coherence and effectiveness of NbS implementation. International advocacy campaigns can raise awareness and mobilize political support for sustainable agricultural practices.

3. Funding and Financial Mechanisms:

- International Funding Programs: Global financial mechanisms, such as the Green Climate Fund (GCF) and the Global Environment Facility (GEF), provide funding for projects that promote NbS. These funds can support research, capacity building, and on-the-ground implementation of sustainable practices.

- Investment Mobilization: International collaboration can attract investment from private sector companies, development banks, and philanthropic organizations. Coordinated efforts can create large-scale, impactful projects that draw significant financial support.

- Solutions: Establishing partnerships between governments, international organizations, and the private sector can mobilize resources for NbS. Developing clear and transparent funding proposals that demonstrate the benefits of NbS can attract international investment and support.

4. Research and Innovation:

- Collaborative Research Initiatives: International research collaborations can address global agricultural challenges and drive innovation in NbS. Joint research projects can explore new technologies, practices, and policies that enhance the sustainability and resilience of agricultural systems.

- Data Sharing and Standardization: Sharing data and developing standardized methodologies for assessing the impacts of NbS can enhance the credibility and comparability of research findings. This can inform policy decisions and support evidence-based practices.

- Solutions: Creating international research consortia and networks can facilitate collaboration and data sharing. Establishing global databases and repositories for NbS-related research can provide accessible resources for researchers and practitioners worldwide.

In conclusion, international collaboration is crucial for scaling up and mainstreaming nature-based solutions in agriculture. By fostering knowledge exchange, policy harmonization, financial support, and research innovation, global cooperation can overcome barriers to adoption and promote sustainable agricultural practices. Embracing international collaboration ensures that NbS can contribute to global sustainability goals and enhance the resilience of agricultural systems worldwide.

Vision for the Future of Sustainable Agriculture

The future of sustainable agriculture lies in a holistic approach that integrates NbS, advanced technologies, and equitable policies to create resilient and productive agricultural systems. This vision encompasses the environmental, social, and economic dimensions of sustainability, ensuring that agriculture can meet the needs of the present without compromising the ability of future generations to do the same.

Integration of Nature-Based Solutions

NbS will be at the heart of sustainable agriculture. These practices, which leverage natural processes to enhance productivity and resilience, will become standard in farming systems worldwide. Agroforestry, cover cropping, conservation tillage, and integrated pest management will be widely adopted, leading to healthier soils, increased biodiversity, and reduced dependence on chemical inputs. By mimicking natural ecosystems, these practices will improve ecosystem services such as pollination, water filtration, and carbon sequestration, contributing to climate mitigation and adaptation.

Technological Advancements

Advanced technologies will play a crucial role in supporting sustainable agriculture. Precision agriculture tools, such as drones, sensors, and satellite imagery, will enable farmers to optimize resource use, monitor crop health, and make data-driven decisions. Biotechnology and genetic engineering will develop crop varieties that are more resilient to pests, diseases, and climate change. These technologies will be integrated with NbS to enhance their effectiveness and scalability. For instance, precision irrigation systems can complement water-efficient crops to reduce water usage, while remote sensing can monitor the success of agroforestry initiatives.

Policy and Economic Support

Supportive policies and economic incentives will be essential for scaling sustainable practices. Governments will implement policies that promote NbS, provide financial support for sustainable agriculture, and create markets for sustainably produced goods. Subsidies will be redirected from high-input conventional agriculture to practices that enhance sustainability. Payment for ecosystem services (PES) programs will reward farmers for providing ecological benefits, such as carbon sequestration and habitat conservation. International agreements and collaborations will harmonize policies and mobilize resources to support global sustainability goals.

Community and Stakeholder Engagement

The future of sustainable agriculture will involve active participation from all stakeholders, including farmers, local communities, NGOs, and the private sector. Farmers will be empowered with knowledge, skills, and resources to adopt sustainable practices. Local communities will be engaged in decision-making processes, ensuring that interventions are culturally appropriate and socially accepted. NGOs and the private sector will play key roles in providing technical support, funding, and market access. Collaboration and partnerships will drive innovation and scale successful practices.

Resilient and Equitable Food Systems

Sustainable agriculture will contribute to resilient and equitable food systems that ensure food security and nutrition for all. Diversified farming systems will reduce vulnerability to shocks and stresses, such as extreme weather events and market fluctuations. Sustainable practices will improve soil health, water availability, and biodiversity, enhancing the resilience of agricultural landscapes. Equitable access to resources, technology, and markets will ensure that all farmers, including smallholders and marginalized groups, benefit from sustainable agriculture. This inclusivity will be critical for achieving social justice and reducing poverty.

Educational and Research Initiatives

Continuous education and research will drive the evolution of sustainable agriculture. Agricultural education institutions will integrate sustainability into their curricula, preparing future generations of farmers and agricultural professionals. Research will focus on developing and refining sustainable practices, understanding the complex interactions within agroecosystems, and exploring new frontiers in agricultural science. Knowledge sharing and capacity building will ensure that innovations reach all farmers and communities.

In conclusion, the vision for the future of sustainable agriculture is one where nature-based solutions, advanced technologies, supportive policies, and active stakeholder engagement converge to create resilient, productive, and equitable food systems. By embracing this holistic approach, we can ensure that agriculture not only meets the demands of a growing population but also preserves the environment and enhances the well-being of all people.

Conclusion

Summary of Key Points

Throughout this book, we have explored the transformative potential of NbS in creating sustainable agricultural systems. We have examined various NbS practices, their benefits, and the challenges associated with their implementation. This summary highlights the key points discussed in each chapter, emphasizing the critical role of NbS in addressing environmental, social, and economic challenges in agriculture.

1. Introduction to NbS and Sustainable Agriculture: We began by defining NbS and their importance in sustainable agriculture. NbS leverage natural processes to enhance productivity, resilience, and environmental health. They include practices such as agroforestry, cover cropping, and integrated pest management, which mimic natural ecosystems and provide multiple ecosystem services.

2. Soil Health and Regeneration: Healthy soils are the foundation of sustainable agriculture. NbS such as cover cropping, crop rotation, composting, and no-till farming improve soil structure, fertility, and microbial activity. These practices enhance soil organic matter, reduce erosion, and increase water retention, leading to more productive and resilient farming systems.

3. Water Management and Conservation: Efficient water management is crucial for sustainable agriculture. NbS such as agroforestry systems, wetland restoration, rainwater harvesting, and constructed wetlands improve water use efficiency, reduce runoff, and enhance groundwater recharge. These practices help mitigate the impacts of droughts and floods, ensuring a stable water supply for crops.

4. Biodiversity Enhancement: Biodiversity is essential for ecosystem stability and agricultural productivity. NbS that enhance

biodiversity, such as hedgerows, buffer strips, pollinator habitats, and integrated pest management, support a diverse range of species and ecological interactions. These practices improve pollination, pest control, and soil health, contributing to more resilient farming systems.

5. Carbon Sequestration and Climate Mitigation: Agriculture plays a significant role in climate change, but NbS offer opportunities for carbon sequestration and climate mitigation. Practices such as afforestation, reforestation, soil carbon sequestration, and perennial cropping systems capture and store carbon, reducing greenhouse gas emissions and enhancing climate resilience.

6. Sustainable Livestock Management: Integrating NbS into livestock management enhances environmental sustainability and animal welfare. Practices such as rotational grazing, silvopasture, and integrated crop-livestock systems improve soil health, reduce greenhouse gas emissions, and provide diverse forage for livestock.

7. Agroforestry Systems: Agroforestry integrates trees and shrubs with crops and livestock, offering multiple benefits such as enhanced biodiversity, improved soil and water quality, and increased climate resilience. Practices such as alley cropping, forest farming, and riparian buffers create multifunctional landscapes that support sustainable agriculture.

8. Urban Agriculture and NbS: Urban agriculture provides fresh produce and green spaces in cities, contributing to food security and environmental health. NbS such as rooftop gardens, vertical farming, and community gardens enhance urban sustainability, reduce the urban heat island effect, and provide recreational and educational opportunities for urban residents.

9. Policy and Economic Incentives: Effective policies and economic incentives are crucial for promoting NbS. Redirecting subsidies, implementing payments for ecosystem services, supporting research and development, and fostering international collaborations can

create an enabling environment for NbS adoption. These measures ensure that sustainable practices are financially viable and widely accepted.

10. Farmer and Community Engagement: Engaging farmers, local communities, and NGOs in the planning and implementation of NbS is essential for their success. Participatory approaches, knowledge sharing, and education empower stakeholders, build community resilience, and ensure that NbS are culturally appropriate and socially accepted.

11. Future Directions and Innovations: The future of sustainable agriculture lies in the integration of NbS with technological innovations and supportive policies. Emerging trends in precision agriculture, remote sensing, artificial intelligence, and biotechnologies enhance the efficiency and scalability of NbS. International collaboration and continuous research are vital for advancing NbS and addressing global agricultural challenges.

The Way Forward for Sustainable Agriculture

The transition to sustainable agriculture requires a concerted effort from all stakeholders, including farmers, policymakers, researchers, and the private sector. The way forward involves several key actions to mainstream NbS and ensure their widespread adoption and success.

1. Policy Reforms and Incentives:

- Supportive Policies: Governments must implement policies that promote sustainable agricultural practices and NbS. This includes revising agricultural subsidies to favor sustainable practices, simplifying regulatory processes, and integrating NbS into national agricultural and environmental policies.

- Economic Incentives: Providing financial incentives, such as payments for ecosystem services (PES), grants, and low-interest loans, can encourage farmers to adopt NbS. Market-based mechanisms, such as carbon markets and eco-certification schemes, can also enhance the economic viability of sustainable practices.

2. Research and Development:

- Innovative Research: Continued investment in research is essential for advancing NbS. This includes studies on soil health, water management, biodiversity, and climate resilience. Collaborative research initiatives and international partnerships can drive innovation and address global agricultural challenges.

- Technology Integration: Developing and integrating advanced technologies, such as precision agriculture tools, drones, and biotechnologies, can enhance the efficiency and scalability of NbS. These technologies can optimize resource use, monitor environmental impacts, and support adaptive management.

3. Capacity Building and Education:

- Training Programs: Comprehensive training programs and extension services are crucial for building the technical skills needed to implement NbS. Governments, NGOs, and research institutions should organize workshops, field days, and farmer-to-farmer learning programs to disseminate knowledge and best practices.

- Educational Curricula: Integrating NbS into agricultural education curricula ensures that future generations of farmers and agricultural professionals are equipped with the knowledge and skills needed for sustainable farming. School gardens and educational programs can also foster an early appreciation for sustainable practices.

4. Community and Stakeholder Engagement:

- Participatory Approaches: Engaging farmers and local communities in the planning, decision-making, and implementation processes ensures that NbS are culturally appropriate and socially accepted. Building trust, demonstrating respect for traditional knowledge, and involving community leaders can facilitate acceptance and adoption.

- Collaborative Partnerships: Forming partnerships with local institutions, NGOs, research organizations, and the private sector can enhance the implementation of NbS. These partnerships bring additional resources, expertise, and legitimacy to NbS projects.

5. Monitoring and Evaluation:

- Robust Systems: Implementing robust monitoring and evaluation systems allows stakeholders to track the progress and impact of NbS. Regular assessments help identify challenges and opportunities for improvement, ensuring that NbS remain effective and beneficial.

- Adaptive Management: Establishing feedback mechanisms and engaging in scenario planning enables communities to respond flexibly to changing conditions and uncertainties. This adaptive approach ensures the long-term success and sustainability of NbS.

Call to Action for Farmers, Policymakers, and Stakeholders

Achieving sustainable agriculture through NbS requires coordinated action from all stakeholders. This call to action outlines specific steps that farmers, policymakers, and other stakeholders can take to support the transition to sustainable agricultural systems.

1. For Farmers:

- Adopt NbS: Farmers are encouraged to adopt NbS practices such as cover cropping, agroforestry, conservation tillage, and integrated pest management. These practices enhance soil health, increase biodiversity, and improve resilience to climate change.

- Participate in Training: Farmers should take advantage of training programs, workshops, and extension services to build their knowledge and skills. Engaging in farmer-to-farmer learning networks can also facilitate the exchange of best practices.

- Collaborate and Innovate: Farmers can collaborate with local communities, NGOs, and research institutions to implement and refine NbS. Experimenting with new practices and technologies can drive innovation and enhance the sustainability of farming systems.

2. For Policymakers:

- Implement Supportive Policies: Policymakers should develop and implement policies that promote NbS and sustainable agricultural practices. This includes revising subsidies, creating economic incentives, and integrating NbS into national and regional agricultural strategies.

- Invest in Research and Development: Governments should allocate funding for research on NbS and sustainable agriculture. Supporting public-private partnerships and international collaborations can drive innovation and address global challenges.

- Enhance Extension Services: Strengthening agricultural extension services is crucial for disseminating knowledge and supporting farmers in adopting NbS. Investing in training programs and capacity building can ensure that extension workers have the skills and resources needed to promote sustainable practices.

3. For NGOs and Civil Society:

- Advocate for NbS: NGOs and civil society organizations should advocate for policies and programs that support NbS. Engaging in policy dialogues and raising awareness about the benefits of NbS can mobilize political and public support.

- Provide Technical Support: NGOs can offer technical assistance and training to farmers and communities, helping them implement NbS effectively. Developing demonstration projects and success stories can illustrate the practical benefits of sustainable practices.

- Foster Community Engagement: NGOs should facilitate participatory approaches that engage local communities in planning and decision-making processes. Building trust and fostering collaboration can enhance the acceptance and sustainability of NbS.

4. For the Private Sector:

- Invest in Sustainable Agriculture: Private sector companies can invest in sustainable agricultural practices and NbS through corporate social responsibility initiatives and impact investing. Supporting research and development, providing financial products, and creating market linkages can enhance the viability of NbS.

- Promote Sustainable Supply Chains: Companies should promote sustainable supply chains by sourcing products from farmers who adopt NbS. Certification schemes and eco-labels can incentivize sustainable practices and create market demand for sustainably produced goods.

- Collaborate with Stakeholders: The private sector can collaborate with governments, NGOs, and research institutions to support the implementation and scaling of NbS. Public-private partnerships can leverage resources and expertise to drive sustainable agricultural development.

Final Thoughts on the Role of NbS in a Sustainable Future

Nature-based solutions offer a promising pathway to sustainable agriculture, addressing the interconnected challenges of environmental degradation, climate change, and food security. By harnessing natural processes, NbS enhance the resilience, productivity, and sustainability of agricultural systems, providing multiple benefits for farmers, communities, and the environment.

1. Environmental Sustainability:

NbS restore and enhance ecosystem services that are vital for sustainable agriculture. Practices such as agroforestry, cover cropping, and wetland restoration improve soil health, conserve water, increase biodiversity, and sequester carbon. These environmental benefits contribute to climate mitigation and adaptation, ensuring that agricultural systems can withstand and recover from environmental shocks and stresses.

2. Social and Economic Benefits:

Sustainable agriculture through NbS supports rural livelihoods, enhances food security, and promotes social well-being. By improving soil fertility, reducing input costs, and increasing yields, NbS provide economic benefits to farmers. Community engagement and participatory approaches ensure that NbS are socially accepted and culturally appropriate, fostering social cohesion and resilience.

3. Global and Local Impact:

The widespread adoption of NbS can have significant global and local impacts. At the global level, NbS contribute to climate change mitigation, biodiversity conservation, and sustainable development goals. Locally, NbS improve the sustainability and resilience of agricultural landscapes, ensuring that communities can thrive in the face of environmental and economic challenges.

4. A Call for Collective Action:

Achieving a sustainable future through NbS requires collective action from all stakeholders. Farmers, policymakers, researchers, NGOs, and the private sector must work together to promote and implement NbS. This collective effort will ensure that sustainable agriculture becomes the norm, not the exception, creating a more resilient and equitable food system for all.

In conclusion, the role of NbS in sustainable agriculture is critical for addressing the pressing challenges of our time. By integrating natural processes into farming systems, we can create resilient, productive, and sustainable agricultural landscapes that benefit both people and the planet. Embracing NbS is not only a necessity but also an opportunity to transform agriculture and secure a sustainable future for generations to come.

www.ingramcontent.com/pod-product-compliance
Lightning Source LLC
Chambersburg PA
CBHW071930210526
45479CB00002B/625